Arnim von Gleich, Christian Pade, Ulrich Petschow, Eugen Pissarskoi

Potentials and Trends in Biomimetics

This study was funded by the Federal Ministry of Education and Research (BMBF)
Funding ID: 16I1551

Federal Ministry
of Education
and Research

Project Management: VDI/VDE Innovation + Technik GmbH, Dr. Marc Bovenschulte

Submitted by:

Universität Bremen

Technological Design
and Development

University of Bremen
Department 4 – Production Engineering
Badgasteiner Str. 1, 28359 Bremen
Prof. Dr. Arnim v. Gleich
Tel: +49 (0) 421 218 2844
E-Mail: mtimmerm@uni-bremen.de

Cooperation partner:

i ö w INSTITUTE FOR ECOLOGICAL ECONOMY RESEARCH

The Institute for Ecological Economy Research (IÖW),
Potsdamer Str. 105
10785 Berlin
Christian Pade
Ulrich Petschow
Eugen Pissarskoi

With the assistance of:
Silke Affinas, Sascha Fechner, Ole Glöckner, Arno Häusler, Laura Spengler,
Sandra Stricker, Sönke Stührmann, Thorsten Wuest

Design:
Andreas Wolter & Ricarda Löser, Weimar

August 2009

Arnim von Gleich, Christian Pade, Ulrich Petschow, Eugen Pissarskoi

Potentials and Trends in Biomimetics

Springer

Prof. Dr. Arnim von Gleich
Universität Bremen
FB 4 Produktionstechnik
FG 10 Technik Gestaltung und
Technologie Entwicklung
28359 Bremen
Germany
mtimmerm@uni-bremen.de

Christian Pade
Institut für Ökologische
Wirtschaftsforschung (IÖW)
Potsdamer Str. 105
10785 Berlin
Germany

Dipl.Vw. Ulrich Petschow
Institut für Ökologische
Wirtschaftsforschung (IÖW)
Potsdamer Str. 105
10785 Berlin
Germany
ulrich.petschow@ioew.de

Eugen Pissarskoi
Institut für Ökologische
Wirtschaftsforschung (IÖW)
Potsdamer Str. 105
10785 Berlin
Germany

ISBN 978-3-642-44406-7 ISBN 978-3-642-05246-0 (eBook)
DOI 10.1007/978-3-642-05246-0
Springer Heidelberg Dordrecht London New York

Cover design: deblik, using an illustration by Marcus Liebich (www.structurate.de)

Printed on acid-free paper

Springer is part of Springer Science+Business Media (www.springer.com)

Foreword

There is a wide consensus about the necessity of sustainable development. There is also a consensus that wide areas of our economy, industry, and technology and the life styles in industrialized countries are not sustainable. Science and technology are widely regarded as (main) causes for this situation. Issues in this context comprise the generally low resource efficiency, an increased and mostly undebated technological power, an increased invasiveness of modern technologies, increasing amounts and diversity of pollutants, and high technological risks.

On the other hand science and technology are also regarded as (main) solution providers towards more sustainability. Thus the question is which type of science and technology is rather a part of the problem, and which type is rather a part of the solution?

'Learning from nature' may give some orientation in this context. Biomimetics and bionics are widely regarded as being a part of the solution. Organisms and ecosystems have learned to solve (technological) problems since the beginning of evolution. In many technological fields they outperform manmade solutions by far. Ecological systems have learned to sustain themselves in dynamic environments. Their achievements are results of an evolutionary optimisation process lasting over millions of years. This is the main reason why biomimetic solutions are widely regarded as not only being ingenious, but also as being ecologically sound, resilient (stable in dynamic environments), and low-risk. These expectations are shared not only by the public and the media, but also by most of the actors in the field itself (Richey 2008). We refer to these sentiments as the 'biomimetic promise' (or the 'biomimetic expectation').

The aim of this study is, to evaluate the potentials and trends in biomimetics and to compare the performance of actors in this field in leading countries. On a more general level, however, this study inspects the evidence for and against the statement that biomimetic solutions can live up to their promise. A promise which, if kept, implies that biomimetics is rather part of the solution than of the problem of unsustainability.

TABLE OF CONTENTS

1 INTRODUCTION

»Nature does not strive to be meaningful, it already is.«

ROBERT WALSER, 1878–1956

1 INTRODUCTION

Biomimetics and bionics are artificial concepts. The term biomimetics is a synthesis of bios (life) and mimesis (to imitate, to mimic). Bionics is a made-up word derived from biology and technics.

The scientific and technological concepts of biomimetics and bionics can be characterized as the attempt of developing technological solutions by learning from nature. Based on a long process of evolutionary optimisation organisms, populations, and eco-systems surpass the achievements in many fields of the previously available technical solutions. Biomimetics turns to "nature's patent bureau" and takes as its role model such organism-based achievements. The public as well as the media hold a relatively positive opinion of biomimetic solutions, which are often a subject of fascination. Biomimetic solutions are generally viewed as ingenious, ecologically appropriate, and low-risk.

The terms biomimetics and bionics are used interchangeable in this publication. We are aware that even further terms could be used (as for example biomimicry or bio-inspired) and that the different terms are often connected with some special meanings in different countries and different disciplines. For example the term bio-inspired is often used in robotics and bionics is in the American scientific literature often used in the sense of augmenting or replacing operations and functions of human extremities through machinery controlled by neural systems (Johnson/Schreuders 2003).

The purpose of this study is to take an in-depth look at the most important trends and potentials of biomimetics as a basis of science and engineering development and innovation and to assess the standing of biomimetics research and development in different countries with the focus on Germany. Based on case studies, literature, patent and research networks analysis, interviews and workshops, statements about the status of biomimetics as a research and development area were worked out. Furthermore a closer look on the validity of the "biomimetic promise" of ecological adaptation and low risk biomimetic solutions, the current trends in biomimetics, as well as the general setting, driving forces and obstacles in the realisation of its potential for innovation were taken.

As with every study, it was necessary to carefully define the subject area and make preliminary decisions with respect to the methodological approach. Among the decisions made regarding the basic approach, the one with the most far-reaching implications for the possible results from this study was, without doubt, the demarcation of the area of investigation.

We chose to look not only at research and development areas and stakeholders who explicitly use terms such as biomimetics or bionics: our goal was also to collect information on those areas that follow a more general biomimetics approach in the sense of "learning from nature" but without making a direct reference to terms such as biomimetics or bionics. Thus the field in an initial step was extended to include large aspects of robotics and prosthetics from the field of medical technology; both modeling, foremost on the example of the human body or human nature. In a further step we also included those areas modeling the fundamental capabilities not only of individual organisms, but also biological systems or more fundamentally, "life forms." The capabilities that characterize life forms is not only self-organization (which we already find in simple forms in chemistry and physics), but rather the ability of self-preservation and self-reproduction in a steady state condition within dynamic environments (homeostasis, adaptability, self-regulation, self-healing). By focusing on such capabilities (many of which bear the prefix "self-") we find ourselves directly on the leading edge of an entire series of highly topical scientific and technological developments, from the material sciences to the hardware and software used in information and communication technologies.

2 TRENDS IN BIOMIMETICS

»Who has not been used to
this world from early childhood
would go mad over it.
The miracle of a simple tree
would destroy him.«

CHRISTIAN MORGENSTERN, 1871–1914

2 TRENDS IN BIOMIMETICS

»Biomimetics« and »learning from nature«

A precise determination of the subject area of this study is anything but trivial. We are working within a narrow understanding of biomimetics and thus concentrating on the research, development, and actors that make direct reference to terms such as *biomimetics* and *bionics*. This viewpoint is then complemented by examinatining broader fields of biomimetics research and development in which these terms do not appear at all or only rarely, but in which a clear, recognizable focus on the role model of nature predominates, or in which an effort is made to find solutions to technical problems – in the broadest sense – by learning from nature. This report concentrates initially on this first area of *biomimetics in the narrower sense*. In the third chapter, with a view to potential technologies, the scope of the investigation is widend to include more general approaches to learning from nature *(biomimetics in a broader sense)*.

2.1 Definition

In the course of the already several-decades-long ongoing debate on the terms *biomimetics* and *bionics* (*Bionik* in German) and the underlying concept of "learning from nature" a number of definitions have been proposed. A selection can be found in Table 1. The listing is in chronological order beginning with the often-cited "first" definition by J. E. Steele from 1960.

History of the terms »biomimetics« and »bionics«

The definitions presented here convey an impression of the difficulties associated with an attempt to pin down the supposedly simple phrase "learning from nature." Problematic is the question of the form and quality this "learning from nature" model has or should have. The definitions suggested range from simple inspiration to the most exact copy possible. At the same time the specific purpose of the learning is a controversial issue which ranges from form-function relationships to systemic (organisational) relationships and from ontogenetic/phylogenetic development processes to the derivation of general guiding principles that can direct technological development.

Definitions are important in order to know what we are talking about. Definitions therefore have a double function: they serve to specify but also to delineate the division between biomimetics and non-biomimetics. Thus it is little surprising that in the biomimetics community controversial debates about such delineations are taking place. The result is that it is not

2

Table 1 — Definitions of biomimetics and bionics from the literature (authors' own compilation)

No.	Author / year	Definition	Source
1	J. E. Steele / 1958–60	"It [bionics] explores systems whose functions are modeled on natural systems, or whose properties resemble those of natural systems, or are analogous to them."	Gérardin (1972, 11)
2	J. E. Steele / 1958–60	"[the] science of systems that work like or in the same manner as or in a similar manner to living systems"	Forth/Schewitzer (1976, 62)
3	L. P. Kraismer / 1967 [initial publication 1962]	"Bionics is thus the science that investigates biological processes and methods with the goal of applying the results to the improvement of older and the creation of newer machines and systems. One could also say that it is the science of systems demonstrating features similar to those of living organisms."	Kraismer (1967, 12)
4	H. Heynert / 1976	"With respect to the present state of development, bionics can be viewed as one of the applied disciplines in the biological sciences with a tendency to integration induced by its objectives, which has as its content the systematic study of life forms for the solution of technical, technological, and architectonic problems; whereby structures and processes serve in their functional relationship in the systems of organisms as a stimulus and pattern, particularly as models for constructions and processes in the various branches of industry and engineering."	Heynert (1976, 37)

→ continuation next page

No.	Author / year	Definition	Source
5	E. Forth & E. Schewitzer / 1976	"Bionics: scientific field of integration, with a technically driven problem focus of heterogeneous scientific disciplines. Their scientific matter is characterized by findings that are acquired from *biological objects,* that embody principles superior to previous technology, and that can lead to a *technical utilisation;* thus / therefore it brings together various disciplines for the solution of specific technical tasks of a varying nature and changing priorities and taps into new types of technical problem-solving approaches." *	Forth / Schewitzer (1976, 58)
6	A. I. Berg / n.d. (possibly 1976 or earlier)	"The task of bionics is to investigate biological objects with the goal of modernizing present technical systems or creating new and more accomplished ones and using the results."	Greguss (1988, 5)
7	E. W. Zerbst / 1987	"In general, bionics can be described by three different groups of definition: (1) It is a science for the planning and constructing of systems whose functions emulate those of biological systems. (2) It is a science for the planning and constructing of systems exhibiting characteristic features of biological systems. (3) It is a science for the planning and constructing of organisational structures that emulate the interrelations of patterns of biological organisation."	Zerbst (1987, 27)
8	VDI-TZ / 1993 **	"Bionics as a scientific discipline looks systematically at the technical conversion and application of constructions, processes, and principles of development in biological systems."	VDI (1993, 10)

* Italics in original; boldface omitted

No.	Author / year	Definition	Source
9	W. Nachtigall / 2002	*The definition from VDI-TZ/1993 (see No. 8) with the following addition:* "Bionics also includes aspects of the interplay of animate and inanimate parts and systems as well as the scientific-technical employment of biological organisation criteria."	Nachtigall (2002, 3)
10	T. Rossmann & C. Tropea / 2005	"Bionics = learning from nature to improve technology"	Rossmann / Tropea (2005a, VII)
11	J. F. V. Vincent et al. / 2006	"Biomimetics (which we here mean to be synonymous with 'biomimesis,' 'biomimicry,' 'bionics,' 'biognosis,' 'biologically inspired design,' and similar words and phrases implying copying or adaptation or derivation from biology) is thus a relatively young study embracing the practical use of mechanisms and functions of biological science in engineering, design, chemistry, electronics, and so on."	J. F. V. Vincent et al. (2006, 471)
12	Y. Bar-Cohen / 2006	"*Bionics* as the term for the field of study involving copying, imitating, and learning from biology ... *Biomimetics* ... [the] term itself is derived from *bios*, meaning life, and *mimesis*, meaning to imitate. This new science represents the study and imitation of nature's methods, designs, and processes. While some of its basic configurations and designs can be copied, many ideas from nature are best adapted when they serve as inspiration for human-made capabilities." [italics in original]	Bar-Cohen (2006, 2)

** The VDI Technology Center in Düsseldorf held a workshop on biomimetics in 1993, at which a dozen of known German experts in the biomimetics community of the time agreed upon a definition of biomimetics.

possible for biomimetics to be defined by any one individual; it will need to gradually crystallize along the course of further developments within the field in order to finally be accepted by a broad majority. Presently, biomimetics still (or perhaps once more) appears to be in a phase of development in which various definitions co-exist.

It is against this background that our own proposed definition must be viewed:

Definition of the term »biomimetics« within the scope of this study

Biomimetics is the attempt to learn from nature; it deals with the development of innovations on the basis of investigation of natural, evolutionarily optimized biological structures, functions, processes, and systems.

Elements of the definition

Within the biomimetics community it became clear that even this definition is not quite adequate for the task of specification and delimitation. Elementary to every definition is, in our opinion, a composition of the three elements that are essential in characterizing biomimetics today: (1) new (technical) possibilities for (2) innovations solving societal problems and/or fulfilling demands and (3) "learning from living nature," or more precisely: learning, in the broadest sense, from "biological research."

Of great importance is therefore the linking of (new) (technological) options with society's problems and needs. Such matching up of possibilities and goals is constitutive for the definition of technology (as the link between means and goals) as well as for innovation (i.e., successful change that fulfills a need). It is a matter of technology and innovation. Biomimetics, specifically, is the source or well-spring at which new (technical) possibilities and solutions are being sought. This source lies less and less in the "direct" observation of nature; it is the biological sciences that deal with the phenomena of animate nature, i.e. the investigation of natural, evolutionarily optimized biological structures, functions, processes, and systems are increasingly serving as a source for innovation.

The definition against the background of present development dynamics

The problem with exact definitions of biomimetics is due in part to the currently rapid rate of dynamic change in the field, as well as the ongoing inclusion of fields in which comparable biomimetics approaches – though not labeled as such – are being pursued, as well as the increasing expansion of biomimetics into neighboring fields of technology and, above all, nanotechnology.

The definition proposed serves as a starting point for the further efforts to describe and circumscribe biomimetics that follows. Thus we shall attempt with the help of the subsequently outlined *three strands of biomimetics development and three levels of learning from nature,* to converge on a conclusive and, above all workable understanding of biomimetics. In connection with the remarks that follow on the relationship of biomimetics and sustainability (the biomimetic promise), our goal is to arrive at a comprehensive as well as reasonably consistent view of the field of biomimetics. In this context, work on a joint guiding principle or mission statement for biomimetic research will soon become much more important than efforts to find a sufficient definition.

2.2 The three main strands of development in biomimetics

Three main strands of bionics development

Greatly simplified, biomimetics development to date can be represented as three successive strands of development in which each following strand has overcome substantial restrictions of its predecessors.

Functional morphology – form and function

First and oldest strand of development: aircraft wings, parachute, hook-and-loop fastener

The first and oldest of these three strands of development focuses on the relationship between biological forms or structures and their functions. The origins are already to be found in pre-scientific observations of nature, which often served as stimuli for technical solutions. Among the most successful innovations in this strand to date are the parachute, the lift-generating aircraft wing, the streamline form and the hook-and-loop fastener (Velcro®).

Form-function dependencies

As long as scientific observations of nature remained in the macroscopic realm, technical implementations within this dimension were able to succeed using the techniques that were thus available; this worked especially well when the desired function was more closely related to its form and less so to the form-giving material. For the lift function of an aircraft wing it is its form, above all, that is decisive. Its technical realisation in a non-biological material does not change that. It is interesting that many of the examples of success in this form-function strand of development derive from the field of fluid dynamics, which leads to a second condition necessary for success. Part of the success of biomimetics in the area of fluid dynamics is due to the fact that the biomimetic approach was capable of compensating for the limitations of mathematical experimental physics. Neither the analytic nor the newer numerical models of fluid mechanics were capable of making calculations or predictions precisely enough to be able to work out optimisations on the board. In the end it was necessary to carry out an empirical experimental trial optimisation process in the fluid-dynamics test chamber (wind tunnel) – and in such trials biological evolution has an enormous lead. As research moves deeper into the relationship between structure and form – from the macroscopic to the microscopic and onto the nano-realm – the more difficult technical implementation problems or "manufacturing

Production-related problems of implementation

Figure 0 — The three main strands of development in biomimetics

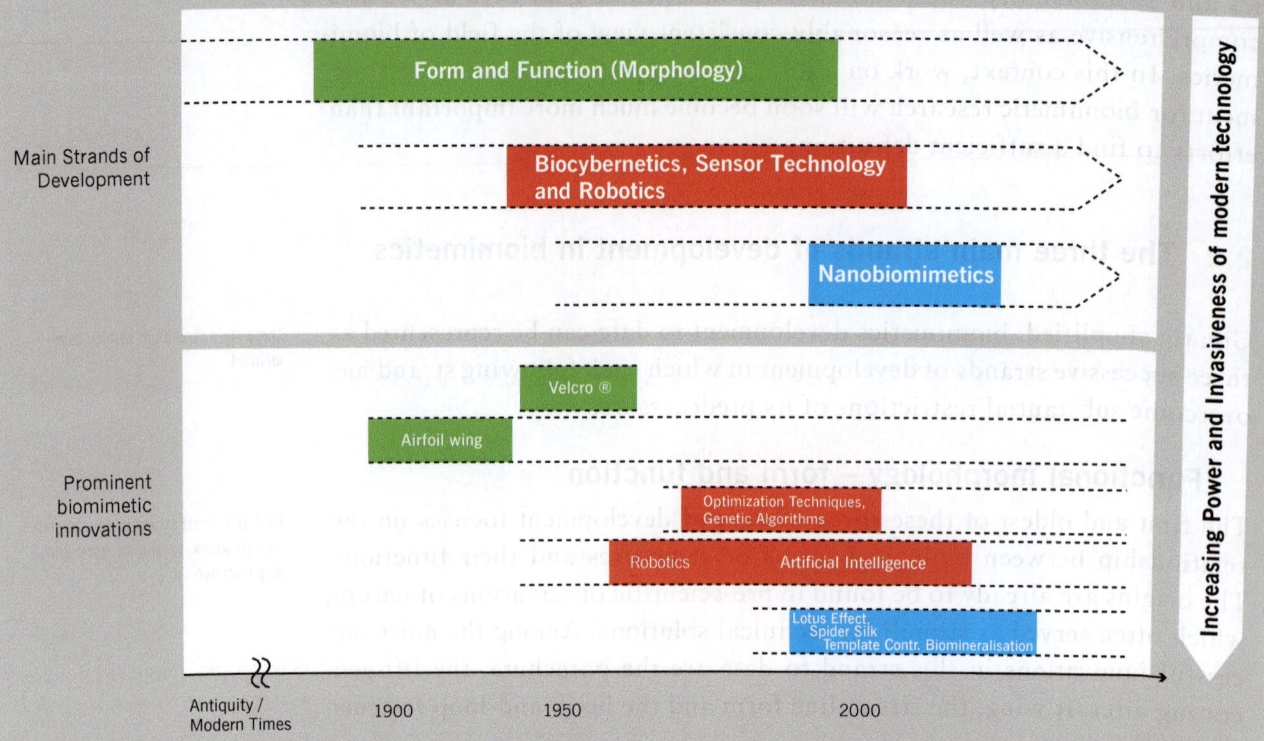

[Source: authors' own diagram]

issues" become. They are among the most significant restrictions today on far-reaching innovation within this field. Highly interesting discoveries for example, in the areas of structured surfaces and functionalized interfaces in biological systems (such as shark skin/riblet film, the lotus effect and anti-fouling) have not yet been technically implemented into production, so that the quality of its results achieved would be comparable to the corresponding natural sources of inspiration. In these examples, it is the quality of the technical realisation that is decisive for the desired functionality and thus for the success of the innovation.

Example lotus leaf

If we take, for example, products on the market today that try to technically implement the surface of the lotus leaf and examine them closely with a scanning electron microscope, it is clear that the technically realized surface is still far from that which could function in the manner of the lotus leaf. Decisive characteristics such as the hierarchical structuring of the papillae and their fine coating of wax crystals have not yet been achieved.

Transformation of the production paradigm

This is likewise true for the hierarchically structured biological materials such as bone, tooth, nacre, and spider silk that are increasingly a focus of biomimetic research. To manufacture such materials or products a fundamental change in our production paradigm is unavoidable. The two previously commanding paradigms of material processing consisted, first of all, of carving a form out of a given block of material (for example, stone or wood); in the second paradigm a material (such as metal or concrete) was first homogenized and then either poured into a mold or forged to form. However, hierarchically structured materials cannot be realized in this manner.

Self-organisation

The solution for manufacturing hierarchically structured materials may be self-organization processes, that means to learn not only from the biological form but also from the process of their formation, i.e. biological development or growth processes. Should this succeed, it would open the door to further desirable properties of such so-called smart materials, for example, the capability of self-healing and the ability to adapt to varying demands. Respective perspectives could open up with the bottom-up nanotechnologies.

Signal and information processing, biocybernetics, sensor technology and robotics

Second main strand of development: robotics and artificial intelligence

While the first and oldest strand of biomimetics development depends on the relationship between form and function, the cybernetic control loop is characteristic of this second strand. Part of this strand are the coining of the term bionics by Steele as well as sensor technology and robotics. This is the strand of development that is commonly referred to as "bionics". In contrast to the first functional-morphological strand, with its development from systematic biology (zoology and botany) to ecology and later on to technical biology and biomimetics, this second strand represents a different, but no less biomimetics-typical developmental logic from the beginning.

Biocybernetics

The fundamental approaches and models of biocybernetics, sensor-physiology and neurophysiology, as well as the ecosystem theory were initially developed in technical areas distant from biology, such as electrical engineering e.g. in resonant circuits, feedback effects, and control circuits, as well as sensors and actuators. Only with their help could important progress in bio-logy have been achieved – particularly in biocybernetics, sensor physiology, neurophysiology and even brain research. This progress in turn positively influenced (not only biomimetic) technical developments in sensor technology, information processing, and robotics. Ultimately in many areas of sensor technology, robotics and information processing up to artificial intelligence (AI), the human mind and body still is the unmatched model.

Artificial intelligence

In the years following the initial euphoria, the area of artificial intelligence has become noticeably quieter. It would seem that in this second strand of biomimetic development things are once again moving forward, if we include current approaches such as decentralized control, parallel computing, self-organizing software, and neuron networks among the biomimetic solutions based on natural models (as well as new actuators such as the pneumatic actuators based on muscles by the company Festo, for example). With the aid of these biomimetic approaches, some of the limitations that have accumulated in the areas of signal and information processing and robotics are being overcome. This second, rather biocybernetic strand of biomimetics seems to be taking on the legacy of artificial intelligence and picking up speed via the fusion of robotics, sensor technology and prosthetics.

Nanobiomimetics – molecular self-organisation and nanotechnology

Third and most recent main development strand: Nanobiomimetics, spider silk and biomineralisation

The third and most recent strand of development in biomimetics is found at the molecular and 'nano' level. This strand also can look back at a longer history (e. g., colloid chemistry, self assembling monolayers). Carried forward by driving forces in the general field of nanotechnology, biomimetic developments in this strand are about to reach a breakthrough (for example, spider silk, biomineralisation, functionalized surfaces, template-controlled crystallisation, neurobiomimetics, nanobiomimetics, etc).

High dynamics in research and development

The nanobiomimetics strand focuses on processes of molecular self-organisation as well as on the (ontogenetic) development of molecules, cells, and tissue, including their reconfiguration (reaction to load) and (self-) healing. With this third and presently extremely dynamic strand, some very promising approaches to solutions are coming up, among these, solutions for the previously mentioned limitations due to "manufacturing hierarchically structured materials" in the first strand of development are rising.

Bottom-up nanotechnologies

Principles of molecular self-organisation, for example template-controlled crystallisation and other bottom-up nanotechnologies, will make possible technical (production) approaches to manufacturing surface textures such as those based on the lotus leaf or shark skin models in the long-term. They also may lead to methods for manufacturing hierarchically structured anisotropic materials based on the model of bones, teeth and plant stems. In a further perspective on development, we can expect "smart materials" capable of reacting to differing loads and, if necessary, even repairing themselves. Presently – and in a foreseeable future – strong dynamics among the three strands are likely to be found in this rather development biology' oriented strand of biomimetics (learning from ontogenetic development processes). Both with respect to the dynamics of the research itself as well as to the possibilities for implementation.

Promising perspectives

For now though, the great innovative successes in this nanobiomimetics strand is yet waiting to happen as most is still at the stage of basic research. But the perspectives are quite interesting and very promising, particularly with the development of hierarchically structured materials – preferably also responsive to load, adaptive and self-healing – as well as functionalized surfaces. Such surface functionalisations can range in effect from reducing resistance or friction to increasing friction, from bonding and adhesive properties (chemical and physical) to self-cleaning, and from biocompatibility to antifouling and biocide effects.

»Biomimetics«

Particularly widespread in this third strand is a phenomenon that could already be seen in the second biocybernetic strand of development. Research and development on molecular self-organisation principles (for example, self-assembling monolayers or SAMs, micelles, artificial biological membranes, functionalized surfaces, template-controlled biomineralisation, etc.) is clearly based on a nature model. It can thus be included in biomimetics, even when researchers themselves do not do so. In the second strand, for example in robotics, labels such as "bio-inspired" are used frequently and relatively carelessly, but in this third strand comparable vocabulary is used rarely – if at all, researchers might fall back on the term "biomimetic." This is particularly true in the English scientific literature, in which this third strand is more often referred to as "biomimetics" (as compared to the use of the term "bionics" in the second strand).

Convergence of the development strands

»Letting things grow«

It would appear as if a trend towards convergence among the three most important strands of biomimetic development exists. In particular, the form and function and nanobiomimetics development strands appear to be merging. In many cases, it is only now with the technical possibilities offered by the third strand, that an extensive realisation and technical implementation of the results derived from the micro- and nanodimension of the functional morphology strand becomes possible. The following brief example of the lotus effect will make this clear. In order to produce a nanostructured surface based on the lotus leaf, it is not only necessary to solve the problems of technical production of the hierarchically structured surface, but also to deal with the significant problems that exist or would exist with the maintenance of this structure over the lifetime of a product. Only when scientists are able to come closer to the biological model and its capabilities of growth and self-repair will they be able to suitably solve these problems. Thus it would appear that in the biomimetics community the feature of autonomous growth of complex (hierarchical) structures will further develop into a guiding principle or model for the development of a biomimetics of the future. This applies above all but possibly not only to the functional morphology and nanobiomimetics strands of biomimetics.

And finally, biomimetics too, may not be able to avoid the present clear trend towards an ongoing fusion of the once separate lines of scientific and technological development. This is currently a prominent topic of discussion under the subject heading "converging technologies" (Roco/Bainbridge 2002, Nordmann 2004). It is quite likely that biomimetics – like nanotechnology – will play an essential role in the course of this consolidation process. The most important candidates that have been put forward for convergence along with nanotechnology, are the information and communication technologies (ICT), the cognitive sciences, robotics, bioengineering and genetic engineering. In all these areas, the concepts of self-organisation, adaptability, self-healing, and self-optimisation (all concepts found in biomimetics) play an important role.

»Converging technologies«

2.3 The three levels of learning from nature

Difficulties of classification

Similarly to the definitions of biomimetics, numerous attempts to internally differentiate and classify the biomimetic field and areas of operations have been publicized. The largest differences in these efforts occur when an attempt is made to classify biomimetics by field of application (e.g. biomimetic civil engineering, biomimetic climatisation, etc.), but then at the same time by technical function (e.g. locomotive biomimetics and sensor biomimetics; see Nachtigall 1998, 19 et sqq.). The problem is that biomimetics approaches are, in principle, possible in almost all areas of science and engineering research and development and even in economics, such that a classification of any kind must be able to accommodate an extremely wide range of science and areas of applications.

Three levels of »learning from nature«

In addition to the subdivision of biomimetics development into its three primary strands of development (functional morphology, signal and information processing and nanobiomimetics), it is also useful to distinguish between three different levels of learning from nature. These three levels consist of: "learning from the results of evolution" (hook-and-loop fasteners, the aircraft wing, etc.), "learning from the process of evolution" (optimisation techniques, evolutionary optimisation strategy (e.g., Evolutions-technik, see Rechenberg/Schwefel), genetic algorithms, etc.), and finally, "learning from the success principles of evolution" (closed loop economy, adaptability, etc.) which is the third and most abstract level. Naturally such an attempt at classifying or structuring must contend with reality, which is substantially more complex; so these strands and levels do not exist in a pure form and working scientists have of course long made use of objects, models, methods, and knowledge acquired from other such strands and levels whenever they seemed useful or promising for their respective research programs.

Learning from the results of evolution

Learning from the results of evolution: shark skin, winglets

The first level, learning from the results of evolution has to date been the most intensively pursued biomimetics. The further technical development of image-making processes, in their spatial as well as temporal resolution (such as the electron or atomic force microscope, slow-motion camera) opened the gates wide on this level, particularly for recording form-function relationships (e.g. the lotus effect and gecko feet using van der Waals' forces, etc.). The ongoing success in modeling the interaction of dynamic bodies (especially their surfaces) and their surrounding media (aero- and hydrodynamics as well as particle dynamics, e.g. the sand skink) may well lead us to expect a strong spurt forward into understanding cause and effect, application, and usage of a targeted generation of microturbulences. This branch of fluid dynamic biomimetics – with recourse to the results of evolutionary optimisation processes lasting millions of years – has thus profited from previous inadequacies of the analytic, mathematic experimental (mechanistic) and also numerical approaches in this field. If, in the future a deeper more functional understanding of this surface-medium interaction can be achieved by numerical methods, a stronger theory-supported and therefore systematically more successful dynamic could develop at this level of learning from nature, which in many areas still largely consists of greatly simplified replication (wings, bulbous bow, shark skin, winglets, etc.).

Evolutionary tested

Regarding these main results of the first level of learning, we may justifiably refer to a certain "validation of technological solutions through evolution". Naturally, for the transfer of this validation into biomimetic innovations, a great deal still depends on the 'process of learning' or the specific knowledge-transfer process, which is based on the fundamental findings. For aero- and hydrodynamic approaches, the recourse to a certain holism is thus also justified, as one more or less takes on the full complexity of the interactions between body and medium; with this, at least the degree of not only theoretical but also practical experimental abstraction remains small. The latter is therefore of importance, because the so-called "noise factors," which can be eliminated relatively successfully within scientific experiments often turn out to be a source of unexpected side effects of technologies (in real life) that later build on these scientific findings (von Gleich 1989 and von Gleich 1998). The experimental and technical power over the object is quite limited in exactly those biomimetic approaches that first and foremost are based on learning from the results of evolution. It is more a matter of fitting in and adapting to the natural circumstances than trying to control them.

Learning from the evolutionary process

Learning from evolution: »soft kill option«, ants' algorithm, evolutionary strategy

Those biomimetic approaches at the second level, where learning from or simulation of the process of evolution (but also the respective ontogenetic adaptation strategies) takes place, have also acquired a comparatively large significance. This would include bioanalogue optimisation methods, for example, evolutionary optimisation strategy (Rechenberg/Schwefel) and evolutionary algorithms, but also swarm intelligence, ant algorithms and the programs developed by Mattheck for component part optimisation (computer-aided optimisation (CAO) and the soft-kill option (SKO)).

Importance of optimisation processes

In attempting to assess the current significance of these processes, their dynamics, and future potential, the results are not so easy to pin down. That may particularly have to do with the fact that the processes themselves as well as the respective application situations (or the respective optimisation problems to be solved) are so heterogeneous that a comparison of the bioanalogue processes to each other as well as a comparison to non-bioanalogue competing optimisation approaches simply does not seem possible.

The best-known bioanalogue methods are apparently in wide application, even though they are in part not always well documented. Currently an exceptionally strong scientific dynamic and particular element of success cannot be seen in any of the bioanalogue or competing methods. Due to the continually increasing complexity in economics and, above all, in the area of logistics, one would have expected more in this connection. Those actors in the field of logistics explain the situation by pointing to a current lack of pressure to solve such problems (or a too-low awareness of the need for such problem-solving). Companies are apparently able to solve their present (optimisation) problems using simpler means and are displaying a certain restraint with respect to the necessary training/adjustment period that more complex methods would require.

Learning from the (success) principles of evolution

Learning from the (success) principles of evolution: solar economy

The third level of learning from nature is based on an abstraction of the – admittedly only somewhat – generalizable principles of evolutionary success. These principles thereby serve a double purpose. They serve as an argument for the validity and scope of the "biomimetic promise" with respect to ecologically more suitable and lower risk technological solutions using biomimetics (see next section). And they can be put into service as guiding principles or models in engineering design, when the goal is to develop ecologically more beneficial, intelligent, adaptive, flexible, robust, and intrinsically safer technical solutions.

Among these principles of success are:

- solar energy and raw material opportunism (use of that which is proximate available)
- modularity, hierarchical structuring, and multi-functionality
- resource efficiency (with a view to limiting factors) and recycling
- resilience (adaptability, diversity, redundancy)
- self-organisation and self-healing
- multi-dimensional optimisation.

2

2.4 The exceptional scientific and technological nature of biomimetics

Biomimetics as a specific form of R&D

Biomimetics, in many respects, should be considered as an exceptional form of research and development. A strong focus on applications links biomimetics to the engineering sciences and their focus on technical solutions, although in contrast to the engineering sciences, which are primarily based on physics and chemistry, biomimetics' foundations derive primarily from biology. Viewed this way, biomimetics is applied "technical biology" (focusing on form-function relations) – but not exclusively. It is also applied neurobiology, applied molecular biology, applied ecology and ecological system theory, applied evolutionary research, and much more.

No traditional engineering science?

On the other hand, one can scarcely call biomimetics an engineering science in the classic sense, since biomimetics in its scientific and technological form, at least, has more in common with biotechnology and computer science. The impulses for biomimetics research and development from the field of biology (new technological possibilities or technology push) are much stronger than those coming from engineering demands (demand pull). Typical of biotechnology, computer engineering, and biomimetics, too, is the equiprimordiality *(Gleichursprünglichkeit)* and particularly close interdependence of technical and scientific development; this form of science has repeatedly been referred to as "techno-science" (see Nordmann 2004).

»techno-science«

However, biomimetics differs in some respects from these "techno-sciences," particularly in its emotional and normative content. The emotional aspect of biomimetics derives its attractiveness from our fascination with (animate) nature. The normative aspect of biomimetics, in its promise of better, more ecological and appropriate solutions, derives from the tested and proven (over millions of years) optimisation of biological models through evolution. The so-called biomimetic promise thus deals with biomimetics' relationship to the topics of risk, ecology, and sustainability.

The "biomimetic promise" – biomimetics and sustainability

Fascinating biomimetics

Initially it can be stated that biomimetics *fascinates* – in a way that is quite different and much stronger than in other areas of research and development. The many (popular) science publications alone make this clear with their numerous wonderful pictures (for example, Blüchel/Malik 2006; Cerman et al. 2005; Kesel 2005; Nachtigall/Blüchel 2003; Benyus 2002, WWF 1991) as well as a whole series of television and radio programs on biomimetics. And not least, biomimetics fascinates young people, for whom engineering might otherwise not be so accessible.[1] Biomimetic topics and solutions enjoy an enormous and enormously positive resonance with the public. Little is known about the reasons behind this fascination and therefore there is much speculation. With great likelihood, the widely held fascination with (animate) nature plays a large role, and also the associated aesthetic and elegance of its many manifestations and modes of action, in which "everything is so perfectly interconnected." As if in confirmation of this hypothesis, almost all (popular science) publications dealing with biomimetics offer up an abundance of especially beautiful and aesthetically pleasing images of nature.

Special quality of biomimetic solutions

Biomimetics also has a significant normative content – and this forms, so to speak, the core of the "biomimetic promise". Biomimetics promises, more or less explicitly, to provide solutions of exceptional quality. This exceptional quality is often justified by reference to the "tested and proven" nature of evolutionary development and an "evolutionary process of optimisation lasting millions of years" (Riechey 2008). The exceptional quality of biomimetic solutions (in addition to the previously mentioned emotional, i.e. particularly aesthetic aspects) is expected to manifest itself in three aspects: (1) in a low degree of risk, (2) in a greater possibility of ecological appropriateness and thus a contribution to sustainability, and (3) in a previously unattainable ingenuity with the respective solution, which mostly (but not only) is justified with reference to a fundamentally multi-dimensional optimisation process. It is presumed – and, it seems not entirely without reason – that solutions which have proven themselves in the competitive process of evolution must possess the respective qualities for success. Even so, the factual justification and respective scope of the validity of this promise needs to be critically questioned.

Reasons for the »biomimetic promise«

But first, let's look at the reasons behind a possible legitimacy of the biomimetic promise. These are closely related to the general evolutionary principles of success, which we already addressed with the third level of

1 In the light of the rapidly growing demand for personnel with scientific and engineering qualifications in the industrial nations in the course of globalisation, and at the same time, a rather sinking – or in any case, certainly not as rapidly growing – interest on behalf of young men and women for an education in the natural sciences and engineering, such corresponding motivation and enthusiasm must be viewed as a valuable (and scarce!) resource, not least from an innovation policy point of view.

learning from nature. The following principles can be formulated as factors for success in biological evolution and as guidelines to a more sustainable technology:

- *Solar energy opportunism*
  Natural processes are essentially based on the usage of solar energy (in the form in which it is available)
- *Raw material opportunism* (usage of what is at hand)
  Biological evolution of organisms is based on comparatively few elements (structure is provided primarily by C, O, H, N, P, Ca, P, S and Si). Organisms primarily use those substances and energy sources that are directly available to them. This does one the one hand radically restrict their possibilities, but it allows them on the other hand to tap into nature's massive energy and bio-geochemical flows and material circles (the latter being a central aspect of sustainability).
- *Resource efficiency and life cycle engineering*
  In those areas where natural resources are among the limiting factors in an ecosystem, efficient utilisation of such scarce resources is an evolutionary advantage. Naturally, a fully closed loop is the theoretical ideal (a closed-loop economy); in nature there are no perfect closed loops either, but when new substances are developed from organisms, as a rule, corresponding catabolic pathways arise in a co-evolutionary process. Organic "waste" is re-used in a cascade-like manner by other organisms, until only humus or mineral components remain.
- *Diversity, redundancy, modularity & multifunctionality*
  The diversity of resources and solutions plays an important productive and stabilizing role in evolution, even when widespread beliefs of a direct relationship between diversity and ecological stability generally represents a crude simplification. Modules such as organs, cells, organelles, and molecules make diversity possible over the basis of a limited set of structures.
- *Multicriteria (multi-dimensional) optimisation in dynamic environments*
  Multicriteria optimisation is perhaps one of the most interesting achievements of the evolutionary processes. Our own technical artifacts are all too often optimized focusing only on a few target functions and are furthermore dependent on well-defined constraints. Organisms in contrast are capable of functioning in turbulent environments. Even in a muddy soup, for example, snails are able to produce their pearlescent nacre. The (biomimetic) optimisation processes, based on the evolutionary processes, offers an approach to multicriteria optimisations.

- *Adaptability, resilience & self-healing*
  Disorders, assaults, etc. are normal in nature. Natural systems cannot limit themselves to only defense or flight. They must learn to manage disorders and assaults without being thrown completely off balance (falling into dramatic system states). Operative terms here are the construction and reinforcement of defense or immune systems; as well as mechanisms for self-healing, resilience, and robustness; and formative goals such as intrinsic safety, etc.
- *Self-organisation*
  The capability of building complex structures (far from thermodynamic equilibrium) is likewise among the most interesting capabilities of organisms and populations and entire evolutionary processes. Opportunities for a biomimetic technological implementation of this principle may already be beginning in the inorganic area with SAMs (self-assembled monolayers) and template-controlled crystallisation. On a long-term perspective, this becomes a question of "letting things grow", of the controlled growth of complex structures and hierarchically structured materials – of which template-controlled crystallization already represents a good example. Such forms of not direct, but only "context-control" make very efficient use of energy, practical monitoring, and control performance.

Validity of the »biomimetic promise« depending on implementation contexts

In reality, biomimetic solutions cannot and do not necessarily need to fulfill all of these aspects at once, but the more that they are met, the greater the legitimacy of the biomimetic promise. Validity, however, can then only be achieved when the respective principles are implemented in the corresponding technology, i.e. in proffered solutions. Thereafter, the respective conditions of use and the application contexts could then also lead to a very different judgment.

Limits of the »biomimetic promise«

A critical examination of the reasons and conditions for the validity of the biomimetic promise thus also shows its clear limits. Here, the following aspects play a key role:

- *A guiding principle is not a label*
  One can treat biomimetics as a guiding principle, whereby the biomimetic promise represents important aspects. But it must be recognized from the beginning, that the pursuit of an overall mission statement or guiding principle for development will not necessarily bring about the corresponding success. "Well meant" – as elsewhere – does not necessarily correlate with "well done." "Biomimetic" should therefore certainly not be misunderstood as being some sort of seal of approval. Such a seal of approval can

only be assigned following a methodically flawless established process of approval; however, such a quality-assurance process does not (yet) exist, and it is doubtful whether such a procedure is even possible.

- *Every technology/innovation must be separately tested and evaluated*
  The following must be considered as elements of a possible hazard or risk: first, the potential hazard that arises out of (new) technical capabilities (in toxicology, for example, this would be the toxicity of a substance), and second, the potential danger that arises in the respective application context (in toxicology this would be the exposure).
- *Ambivalence of opportunism (as an enormous form of self-restraint)*
  The use of solar energy as an energy source and the use of the materials immediately at hand (tapping into nature's massive energy and bio-geochemical flows and material circles) certainly increase the probability of a sustainable solution, but at the same time, they drastically limit the possibilities of human creativity and the art of engineering.
- *The more radically new the structure, the more tenuous its claim on the validation by evolution.*
  The stronger the constructive (synthetic) element of a biomimetic approach or innovation is, the more the argument of being proven by evolution becomes invalid. The more profound the depth of intervention and the more powerful the technology is, the more likely it is that there will be relevant side effects and consequential effects (the greater the reach through time and space of the resulting cause-and-effect chains).

2.5 Tentative conclusion about trends in biomimetics

Heterogeneity and dynamics of biomimetics

Even if one only concentrates on biomimetics in the narrow sense, it is not easy to precisely describe this field; this is due to its heterogeneity, but also the dynamics of biomimetics research and developments. Looking at the trends in biomimetics i.e., the dynamics of the field, a differentiation between the three main developmental biomimetics strands can be made. The first, "functional morphological", strand is the oldest and dedicated above all to the form/structure-function relationships. The second strand focuses more on the biological forms of signal and information processing and introduced successful technical implementations; particularly in the areas of biocybernetics, sensorics, and robotics. The third and youngest strand of development in biomimetics is a result of progress in the area of nanotech-

Three strands of development in biomimetics

nology and draws on, among other areas, molecular self-organisation processes. In our opinion, taking a mid to long-term perspective, the greatest potential for biomimetics may well lie in this area. A significant reason for this is that this third strand is capable of overcoming significant restrictions of the first two strands.

Additionally, latest observations suggest that the convergence of these three strands is becoming ever greater, to some extent they may have already merged. This convergence has its counterpart in other areas of science and technology and is frequently referred to as "converging technologies."

»converging technologies«

In addition to the three strands in the development dynamic, the core of the biomimetic basic idea can also be broken down further. We distinguish between three levels of "learning from nature," where the question arises as to the conceptual source of the respective biomimetic knowledge-transfer processes. In the case of "learning from nature's findings," it is the structures and mechanisms of living systems that are found in nature and described in biology that are used as models for technical products and processes. At the second level, it is neither the findings nor the results of evolution but rather the evolutionary process itself that is the object and starting point of the knowledge transfer processes in biomimetics (evolutionary optimisation, genetic algorithms). There is also an effort underway in biomimetics to distill out the general principles of the evolutionary success and the structure and functionality of natural systems (for example, resource efficiency, opportunism and adaptability); this can be considered the third level of learning from nature.

Three levels of »learning from nature«

Finally, biomimetics clearly stands out from other areas of science and technology with two more characteristics: a) the fascination it exerts on many, inside as well as outside the sciences, and b) the normative content which it more or less implicitly transports. The fascination may well be attributed to the fact that biomimetics, regardless of the form in which it is pursued always takes living nature with its positive attributes as its starting point and, in a way, as its goal. The intent is to emulate nature, in its fascinating and endless variety, not to manipulate it or fight against it. The normative content shows itself, particularly in the popular reference to the evolutionary seal of approval, i.e., that the systems (organisms, etc.) of living nature have undergone an evolutionary process lasting millions of years during the course of which they have in many respects become "optimized." This optimisation took place not only in the respective individual systems, but also and foremost in the interactions between the system and its environment and the resulting effects. The latter can be associated with a certain degree of ecological "suitability" or appropriateness which, against the backdrop of the ongoing debates on sustainable technology (and development) lets biomimetics appear as a very promising alternative.

The »biomimetic promise«

3 TECHNOLOGICAL ASPECTS OF LEARNING FROM NATURE

»We must not believe that all miracles of nature can only be found in other countries and on other continents. They are everythere. But we do not appreciate those ones surrounding us, because we have been watching them each and every day from childhood on.«

JOHANN PETER HEBEL, 1760–1826

3 TECHNOLOGICAL ASPECTS OF LEARNING FROM NATURE

In this chapter we address the current trends and the technical prospects of biomimetics as a field of research and development. Two general approaches were chosen – one from within the field of biomimetics, and one from outside. This involved, first of all, an analysis of general technology foresight studies and a correlation of the biomimetics R&D fields and trends towards the particularly promising fields of technology. Then, by means of case studies, a close look is taken at the medium-term potentials of particularly promising areas of biomimetics.

Evidence and statements on the prospects and trends in biomimetics thus derive from three sources: a) analyses of general technology foresight studies and studies from other R&D fields, b) an analysis of the scientific and paradigmatic dynamics of biomimetics, and c) a look at the situation concerning society's demands and needs, in other words, the challenges facing biomimetics. Essential information about the demand and general technological trends was obtained from technology forecasts; essential information with respect to scientific dynamics and possible contributions to solutions focusing on urgent societal challenges were obtained by case studies.

See chapter 3.1

We begin with the evaluation of the technology foresight studies; this was done to determine the extent to which they can offer clues to the future significance of biomimetics or research approaches that try to mimic nature in the broadest sense. The results are summarized in section 3.1. It was possible to identify a series of scientific areas in which the potential (expected and desired) results presume the application of biomimetic approaches or such approaches appear promising.

See chapter 3.2

In a second stage, case studies were completed for some of these scientific fields. These relied primarily on analyses of literature and expert interviews to determine the dominate trends, the challenges for each research subject, and the extent to which the participants are focus on "learning from nature" to cope with these challenges. The results of the case studies are summarized in section 3.2.

See chapter 3.3

Section 3.3 contains the synthesis of the three approaches with statements on particularly promising trends and prospects in the narrower domain of biomimetics, i.e. in situations where the stakeholders are explicitly identified with biomimetic approaches, as well as in the larger context in which, for whatever reasons, this is not the case.

3.1 Biomimetics and its technology foresight

Technology foresight to identify promising R&D approaches

Technology foresight has established itself as an approach and tool in the strategic management departments of larger enterprises, in industrial federations, and among funding institutions (VDI 2004, VDI 2006); the objectives differ accordingly. The authors of such foresight studies try to identify technologies and areas of science that promise significant social or corporate benefits, they look for challenges where the need for a technological solution appears to be particularly urgent; however, sometimes they also attempt to represent their own approaches, expertise and current projects as being especially dynamic and promising.[2]

National technology prognoses

We first looked at nationally funded studies intended to aid in setting priorities in the funding and promotion of research, technology and innovation. These included technology foresight studies for Japan (NISTEP 2005), the USA (Popper et al. 1998), France (MEFI 2006), and China (RGTF 2005).[3] The Chinese foresight study was included because literature analyses show a large share of Chinese researchers worldwide in publications on biomimetics (see chapter 6 on the international R&D landscape). The question arose as to whether this high relevance of biomimetics was to be found also in Chinese technology foresight studies.

See chapter 6

Supranational technology prognoses

Furthermore, future development studies, commissioned by supranational stakeholders – in particular the United Nations and the OECD – were considered.[4] In this context, the study on key technologies commissioned by the EU commission with its more general technology orientation is of special interest,[5] because the support of so-called key technologies is expected to make a particular contribution of strategic importance to innovation and international competitiveness (see Key Technologies Expert Group 2005, 25ff.)

Finally, a number of studies on future development commissioned by large enterprises with varying priorities are publicly available. The Shell study on the future of energy supplies (see Shell 2005), the BMW study on the future of mobility (see ifmo 2006), and again with a broader focus on technology foresight, the Siemens report "Horizons 2020: A thought-provoking look at the future" (TNS 2004) appear particularly relevant here.

In the course of assessing these foresight studies, the intent was not only to look at explicit references to biomimetics is wanted, but also consider to which of the prioritized technical challenges, could biomimetic

2 See Salo and Cuhls (2003) for an introduction to the difficulties and challenges of technology forecasts.

3 An overview of the available international technology foresight studies can be found in VDI-Technologiezentrum 2004 and VDI-Technologiezentrum 2006.

4 See Glenn/Gordon (2006), OECD (2006), IEA (2006)

5 In the study from the International Energy Agency the focus is, as expected, on the development of energy technologies and energy supplies until 2050. The UN studies concentrate on the social development and the future of sustainable development.

Investigation of science and implementation fields

approaches contribute with possible solutions. In this context, it is first necessary to clarify the understanding of biomimetics for the purposes of this study. For the purpose of structuring the current research on biomimetics we relied on the classifications found in the biomimetics textbook by Professor W. Nachtigall, whose (thoroughly successful) efforts to present an overview of developments in this field of research (Nachtigall 2002, VII) can be acknowledged as the most comprehensive and detailed publication available on the state of development of biomimetics R&D in German language. At the same time, we also made use of the classifications of the German BioKoN research network.[6] The two arrangements are largely compatible, but it is clear that BioKoN, (with terminology such as "surfaces and interfaces" and "biomechatronics and MEMS") makes a greater effort in the direction of compatibility with modern technological development (see Table 2) than can be said about Nachtigall (2002).

Table 2 — Structural classification of R&D fields of biomimetics according to BioKoN and Nachtigall

BioKoN		Nachtigall
B1	Architecture, design	Styling and design; construction and air-conditioning
B2	Light-weight construction, materials	Materials and structures
B3	Surfaces and interfaces – structures and functions	
B4	Fluid dynamics, flying, swimming, robotics, DAMS	Robotics and locomotion
B5	Biomechatronics, biomedical technology, microelectromechanical systems (MEMS), actuatorics, robotics	Anthropomedical and biomedical technology
B6	Sensorics, information processing, communication	Sensors and neuronal control
B7	Biomimetic optimisation methods	Evolution and optimisation; Systematics and organisation
		Structures and devices (including nanotechnology)
		Procedures and routines (renewable types of energy)

[Source: Authors' own compilation based on Nachtigall 2002 and www.biokon.de]

Hints at »learning from nature«

To answer the question whether biomimetic approaches come into application or consideration in those fields singled out in the foresight studies, one can – in addition to considering explicit references to *biomimetics/bionics,* and a comparison with the potentials of biomimetics outlined in the table – look to whether an intuitive (heuristic) preconception of each

6 www.biokon.net

technology or its potential applications can be leveled with the concept of learning from nature or whether at least references to it can be drawn. For example, the R&D field of artificial photosynthesis can be intuitively associated with the imitation of organisms (here: plants and bacteria practising photosynthesis).

National technology foresight studies

Qualitative and methodic differences of national technology prognoses

The methodological basis and prerequisites of the studies differed greatly. The Japanese and Chinese investigations were based on the Delphi method, in which selected experts from science, society, and politics assessed possible future developments in several rounds of consultation. The results of these studies are as a rule, more or less quality-controlled surveys of possible matches between the challenges and the respective technological problem-solving approaches (topics).[7] These surveys can be classified according to various criteria of social and economic significance. The French study (MEFI 2006) additionally set up a Web site offering visitors the opportunity to propose technologies having a potential for the future. The results gathered from these studies are based on assessments by expert groups from business, politics and science. They worked in eight subject groups and addressed the challenges of their respective topics and discussed the results in four workshops in plenary sessions. The US study was based on interviews with selected persons from various companies, among them members of technological boards, CEOs, and the heads of planning and development departments.

Often no explicit references to biomimetics...

...rather, many references to "learning from nature"

To begin, it can be stated that those technological applications identified as the most important in the foresight studies often have no explicit reference to biomimetics. In the opinion of the majority of experts, biomimetics does not play an outstanding role. However, references to biomimetic approaches do become apparent when we look beyon the explicit mention of terms such as biomimetics and bionics (and their variants) to a broader understanding of biomimetics. This would include such concepts and terminology as *self-healing, self-organisation, intelligent, smart, biologically inspired,* etc. Furthermore, a broader, more superficial examination of the foresight studies reveals a slightly different picture regarding the relevance of biomimetics than a more detailed analysis does, that also specifically considers the subregions of larger fields of technology. For example, applications from the field of robotics are regarded as having a high relevance in one of the foresight studies. Since robotics, however, is based

7 Examples for such topics from the Japanese study include: "A risk-management system that utilizes disaster observation satellites, communications satellites, GPS, unmanned aircraft, and so on to observe disasters, understand situations after disastres occur, and respond swiftly." Secondly: "Technology to forecast the timing and scale of volcanic eruptions by observing and assessing in real time magma conditions inside volcanoes that are likely to erupt."

Table 3 — Biomimetic research fields with reference to promising technologies from the technology foresight studies

Field of biomimetic research	Technology fields with affinity to biomimetic approaches
Materials and structures, surfaces	new materials using nanotechnologies (NISTEP 2005) sustainable materials (RGTF 2005, Popper et al. 1998, MEFI 2006) biomaterials (Popper et al. 2001) complex polymers (Popper et al. 1998) new materials due to an understanding of self-arrangement in natural processes (Popper et al. 1998, MEFI 2006) intelligent materials (MEFI 2006)
Robotics and locomotion	intelligent robots (Popper et al. 1998, NISTEP 2005) reliable, safe machine technologies (RGTF 2005) adaptable machines (Popper et al. 1998)
Anthropomedical and biomedical technology	tissue production (NISTEP 2005) biological pharmaceutics (NISTEP 2005)
Sensors and neuronal control	nanosensors (NISTEP 2005) intelligent sensors (NISTEP 2005, RGTF 2005, MEFI 2006) biosensors (Popper et al. 1998)
Information processing and communication (study group at BioKoN)	bioelectronics, biocomputers (NISTEP 2005, RGTF 2005) electronic technologies using biological knowledge: nanotechnologies; molecular self-organisation as a model for the construction of complex systems (NISTEP 2005) new network architecture, smart networks (RGTF 2005) better man-machine communication (Popper et al. 1998) computers adapting themselves to man (MEFI 2006) autonomic computing (MEFI 2006)
Evolution and optimisation; systematics and organisation	
Structures and devices (including nanotechnology)	self-organising technologies (in various applications)

[Source: authors' own compilation]

on the human model in extensive areas of its hardware and software development (see Nachtigall 2002, Cerman et al. 2005), we can assume that biomimetic approaches are highly relevant here, even if this is not explicit.

Intersection of biomimetic R&D and promising technologies

The results of our search for possible intersections between biomimetic R&D fields and the most promising technology fields as derived from the technology foresight studies are presented in two tables. In Table 3, technologies and applications from the studies were assigned to categories of biomimetic research based on those defined by Nachtigall (2002) and the

BioKoN network (www.biokon.net). Each assigned technology field had to meet three criteria:

1. a high degree of relevance within the foresight studies;
2. it can be assigned to one of the categories of biomimetics research;
3. the intuitive preconception of the technology or its application allows it to be related to the idea of mimicking nature.

Thus, beginning with the biomimetic research fields, connections have been drawn to the technology fields derived from the technology foresight studies.

Specifics of the »Rand-study«

The authors of the Rand study (Popper et al. 1998) – which was intended to identify technologies promising technological breakthroughs on a long term – chose the reverse approach. They did not present those queries with a prepared selection of precise technological tasks (as in the Delphi studies), but asked their respondents to name technologies (e.g. R&D fields) expected to produce long-run promising breakthroughs.

Fields of technology with potential relation to biomimetics

The respondents named a number of technological fields in which connections to areas of biomimetic research and development can be presumed: "computer intelligence by means of evolutionary programming and genetic algorithms," "artificial intelligence," "biomaterials and biologically produced materials," "emulation of natural materials using nanotechnologies," and "imitation of biological production methods to creating a more environmentally friendly production processes" (Popper et al. 1998, 60). One of the persons interviewed put it very concisely: "Biosensors and the marriage of organic materials to inorganics in general have no connection to the work we are presently doing but would seem to be of general importance. The marriage of biology to engineering could lead to development of such things as organic light-emitting diode readouts. Various forms of self-assembly systems that are characteristic of biological systems could then be applied to materials processing or the assembly of mechanical systems" (ibid. 35).

Various interpretations

However, one should be careful in interpreting such statements. For instance, the term *intelligence* seems to be ostensibly synonymous with *biomimetics* since intelligence, as a rule, is only assigned to higher-level creatures. But the extent to which development aiming at *intelligent systems* really tries to emulate human (or animal) intelligence – or is even capable of this – is highly uncertain. On the face of it, it is not wrong to assume that research work on intelligent systems takes as its model intelligence from the natural world. The extent to which such research can be considered as being biomimetic research must be determined in each individual case, and in the end it also depends on the specific interpretation of

See chapter 3.2

biomimetics being applied. This difficulty is addressed in more detail in the case studies (see section 3.2).

Supranational technology foresight study: EU-Foresight

EU-study on key technologies

The EU-foresight study chose a very technology-oriented approach to so-called key technologies. The European Commission had entrusted the department of Science and Technology Foresight of its research directorate with the task of evaluating newly developing research fields and assessing their relevance for the EU (Key Technologies Expert Group 2005).[8] A study was prepared for each of the fifteen branches in order to obtain an overview of scientific challenges and the current state within the respective fields. Subsequently, the results of the individual studies were discussed at a conference and general conclusions for science policy were agreed upon. These were then published in an additional report which also contained an analysis of the strengths, weaknesses, opportunities, and threats (SWOT) of the individual key technologies.

Recommendations for a long-term agenda

Among the policy recommendations is an action plan with calling for a long-term vision for research. It includes the call for a change to a bio-economy, i.e. an economy based on renewable resources and a higher resource efficiency and thus lower negative environmental impact. The researchers formulated recommendations for a long-term research agenda over the basis of the results of the individual studies.

Explicit reference to biomimetics

With respect to biomimetics, a recommendation by the Key Technologies Expert Group (2005) is of particular interest. It proposes to provide sound financial funding for those scientific projects that cannot clearly be assigned to one of the traditional disciplines – explicitly mentioned are the cognitive sciences, the complexity and social sciences, and *biomimetics* (ibid. 8, 60). But the report does not indicate any reasons why biomimetics deserve support, nor how the authors define biomimetics. This becomes clearer only when the individual studies on nanotechnology (Saxl 2005) and complexity science (Priami 2005) are taken into consideration.

Nanotechnology and science of complexity

Priami (2005a) assumes that coping with complexity is among the central challenges of research today. Until now, we have been trying to cope with complex situations by reducing them to their constituent parts in an attempt to try to understand the relationship between them, Priami refers to this as a "reductionist" approach. For today's challenges, however, the reductionist approach is no longer appropriate, instead there is a need for a systemic treatment of phenomena. The author explains this as follows:

8 The fifteen areas were: biotechnology, nanotechnology, information technologies, communication technologies, transport technologies, energy technologies, environmental research, social sciences and humanities, manufacturing and materials technologies, health research, agricultural research, cognitive sciences, safety technologies, complexity research and systemic research in the services sector. *Source:* http://cordis.europa.eu/foresight/kte_expert_group_2005.htm

"The basic idea is to view the systems as something that has its own behavior not obtained simply by gluing [together] the behavior of the systems' components of which we already have all the information" (Priami 2005, 2); a systemic approach is instead taken in systems biology. Priami (2005) assumes that in (animate) nature there were starting points and models to be found for an appropriate approach to dealing with complexity. The author also sees a good approach to "mastering complexity" in the convergence of technologies (ibid., 5); biomimetics plays here an important role in his opinion: "Also industries in the IT sector are investing in the study of the complexity to try to mimic how nature handle[s] complexity to devise new product[s] that scale one order of magnitude in the size of the problems that they can handle" (ibid., 5). Thus, Priami: "The biomimetic approach seems to be very promising in the next future" (ibid., 11).

Nano constructions applying lego-princinple

Saxl (2005), too, assumes that biomimetics will first make possible many significant technological approaches in the nanosciences (see pp. 10, 16, 30). A mechanistically based macrostructure construction out of nanoparticles modeled on bricks or 'Lego' building blocks has proven to be extremely complicated or even impossible (ibid. 10). The author sees a way out in the emulation of nature, but it is first necessary, he suggests, to better understand how natural organisms and processes function at the nano-level. In his opinion, this could best be achieved by way of interdisciplinary research teams. He therefore concludes:

> *"If, as is postulated, biomimetics is the route to disruptive new products and processes based on nanotechnology and other disciplines, then this ability to work in teams is of great importance to Europe if it is to succeed in researching and applying biomimetics-inspired technologies."*
>
> (SAXL 2005, 16)

Biomimetics as key technology?

In their general conclusions, the Key Technologies Expert Group (2005) mentioned a number of research areas pursuing objectives similar to those formulated for biomimetic approaches. The degree of importance assigned to biomimetics is therefore remarkable. Biomimetics, as a specific focus on the emulation of problem-solving approaches from living nature, is explicitly included in the key technologies of the future; contributions to suitable handling of complexity and innovation in the nano-range are expected in particular. For these reasons, the Key Technologies Group recommended the acknowledgement of biomimetics as an extremely promising interdisciplinary approach as well as funding for this line of research.

Technology foresight studies by companies

Large companies active in international markets must be able to identify social trends as early as possible in order to be able to respond to threats and recognize opportunities in a timely manner. Here we focus on a broad study commissioned by Siemens and carried out by TNS Infratest (TNS 2004).

Depiction of future scenarios

Developments which could have an influence on the world of tomorrow are presented in this study in the form of social scenarios. They are based on assumptions of how technologies, will develop in the coming – in this case, sixteen – years. Such technological projections were prepared by the corporate technology department of Siemens and were published in their own corporate magazine "Pictures of the Future" (TNS 2004, 151). The image of future technologies resulting from these projections was summarized by the authors of the study in a chapter entitled "Technology Foresight." In their enumeration of technologies with great potential for the future, the authors mentioned biomimetics as well as bionics. However, the interpretation of bionics held by the authors of the TNS study seems to differ from this study in certain aspects.[9]

Many indirect references to biomimetics

Indeed, the study mentions a number of other approaches which attempt to emulate or use models of nature without classifying them as being either biomimetic or bionic. There are mentions in the field of medicine[10] (including biochips, biosensors, tissue engineering), in materials technology and nanotechnology, and in the interlinking of biotechnology, nanotechnology, and information technology.

Materials tailoring

In the material sciences, so-called "materials tailoring," the development of materials having specially tailored properties, is emphasized. This approach entails interdisciplinary development teams consisting of material science researchers, physicists, chemists, engineers, computer scientists, and interestingly, also biologists (TNS 2004, 210). A further main trend with clear connections to biomimetics is bioengineering. The idea for this branch of research is described by the authors as follows: "Taking nature as our model, we use biological materials, concepts and processes to provide revolutionary solutions in technology and healthcare" (ibid., 216). The main trend, "adaptive structures with intelligent materials," also explicitly refers to models from nature: "Taking a cue from nature, materials systems which use properties like status monitoring to anticipate breakdowns repair themselves, and adapt their properties to suit their surround-

Bioengineering

Adaptive structures

9 With respect to biomimetics, they write, for example: "Today, modifications of biological materials – for example, conventional processes like leather tanning and charcoal production – are called 'biomimetics'. [...] Biomimetics also includes the replication of natural materials." (TNS 2004, 217). At least, as an example, the artificial production of spider silk is put forward (ibid. 217). Possibly following the example of the international linguistic use, bionics is mentioned in connection with information processing: "Information bionics – the use of biomolecules as calculating machines" (ibid. 218).

10 The authors of the study call the area "health care."

ings have been developed" (ibid., 218). This research approach deals with "smart," "shape-memory," and "self-healing materials." But much more than that, it is also about the production of fault-tolerant systems: "The application of biological principles is driving the successful development of fault-tolerant architectures and error-correcting algorithms" (ibid., 238); "a variety of nanoscale analytical devices that intimately combine biological and physical elements have been created to detect and measure a variety of physiological stimuli" (ibid., 245).

Converging of bio-, nano- and information technologies

The authors postulate that some major technological trends will only be realized through the interplay or convergence of biotechnology, nanotechnology and information technology. Ubiquitous, invisible and intelligent information technology, quantum mechanics–based calculators, and the use of photonics in everyday applications are explicitly mentioned. In some of the research areas, the Siemens study describes trends that have no direct reference to the emulation of natural models, but that nonetheless target the realisation of properties such as intelligence and self-organisation. Among the major trends in the field are "production and automation systems": "intelligent buildings," "self-tutoring algorithms," and "lab-on-a-chip" (ibid., 152). In the area of information and communication technologies, the authors see important trends in "ubiquitous computing" and "self-organising networks".

»Learning from nature« highly appreciated

In short, it is clear that in the opinion of the authors of the Siemens study, approaches that try to emulate nature will clearly play a significant role in to future technological development. In particular, they emphasise the inspiration of (living) nature in the field of material sciences. But they also point out a further trend: the authors see promising applications for the future in the convergence and interlinking of biology, chemistry, physics, IT and nanotechnology.

3.2 Results of the case studies

The statements from the studies on technology trends and so-called key technologies presented in the previous section are fairly abstract in nature. For more concise insights and information on biomimetic trends and prospects, it is necessary to go into more detail by means of case studies and in-depth studies. In this step, the focus is not on general technology trends from which references to biomimetics can be then formulated, but rather on the biomimetic developments as such. The case studies are intended to demonstrate the broadness of the biomimetics field by means of nine – in part very diverse – areas from which connections to biomimetic research are then drawn.

Nine case studies

Selection of case studies

The selection of the case studies was based on several aspects and criteria. First, a rather broad definition of biomimetics was assumed. Those fields in which animate nature serves as model were included, even when terms such as *biomimetics* and *bionics* were not explicitly mentioned. This strategy was a consequence of the outcome of our analyses of the technology foresight studies. In fact, these investigations showed that in the foresight studies terms such as *biomimetics* or *bionics* appear relatively seldom; much more often approaches and concepts are described that (can) directly or indirectly rely on nature as a model for technical solutions. These approaches and concepts, generally, use such terms as *biologically inspired*, or *self-healing*, as well as *self-organizing, intelligent* and *smart*. In order to also include those approaches that are in fact biomimetic but are not classified as such, we also specifically searched for alternative biomimetic terms in our investigations as a part of the case studies.

Methodical difficulties with the wide biomimetics term »learning from nature«

Of course, this can be a new source of errors – not all projects whose aims are "intelligent" or "self-healing" technical systems necessarily model themselves on the intelligence or self-repair mechanisms of biological systems. Thus, a point of departure which also includes implicit biomimetic approaches will inevitably lead to fuzzy results; nonetheless, it is our opinion that in this way a more realistic image of the diversity and scope of biomimetic research and development could be achieved, than it would have been possible by strictly limiting ourselves to explicitly biomimetic approaches.

Focus on high-tech biomimetics

In the selection, we also incorporated the results of the technology foresight studies as well as our own initial understanding of the dynamics of technology. Research-intensive and technology-intensive developments (i.e. "high-tech-biomimetics") were favored. Here too, there is room for potential error. After all, such successful biomimetic innovations of the past as the hook-and-loop fastener or lift-producing aircraft wing would not really be considered high-tech biomimetics today. In our selection of case studies and in-depth studies, we worked under the assumption that limiting our observations to the earlier main strand of biomimetics development, i.e. to the close connection between a form-function relationship found in nature and a subsequent manufacturable solution that is then relatively simple to realize at the macroscopic level, would inappropriately limit the future prospects of biomimetics. On the contrary, indications are that in this biomimetics strand too, the knowledge as well as the technical production solutions are increasingly shifting to the micro-(system) technological or even nanotechnological area, i.e. into the third strand.

Smart technologies

Another major trend, derived from the technology foresight studies, is the growing significance of smart technologies, i.e. of the increasing intelligence, adaptivity and self-organisation of technical systems. As representative of a group within biomimetics that is internationally growing in importance and brings its own respective (self)-interpretation of biomimetics, Bar-Cohen should be mentioned (see Bar-Cohen 2003 and 2005).

Table 4 again lists those technological trends that were taken into consideration in the selection of case studies.

Consideration of literature- and patent analyses

Table 4 — Especially dynamic technology fields linked to biomimetics

ICT and electronics	• self-organizing media • processing of large data volumes • man-machine interfaces • electronic technologies applying biological knowledge (mechatronics, MEMS, molecular self-organisation as a model for the construction of complex systems)
Medicine	• prosthetics • biocompatible materials • tissue engineering
Nanotechnology	• self-organizing technologies (in numerous applications) • biocomputers • tissue engineering
Materials	• biologically inspired materials • biocompatible materials
Manufacture and production methods	• adaptive production facilities • automation technology

[Source: authors' own compilation (based on results of analyses of technology foresight studies)]

The results of the literature and patent analyses (see chapter 4) provided additional support for this rationale in selecting the case studies. Certainly, general methodic criteria, such as accessibility to the respective field, comprehensibility, and the limitability of the assignment also played a role in the selection of case studies.

See chapter 4

- Short-term case studies were then carried out in the following areas:
- Sensorics (sensor technology)
- Robotics
- Information and communication technology (ICT)
- Biologically-inspired materials
- Bioceramics
- Prosthetics
- Tissue engineering
- Bottom-up nanotechnology (nano2bio)
- Converging technologies

Established...

Robotics, sensorics, prosthetics, bioceramics, and biologically-inspired materials were chosen for study because, among other things, they are well-established fields within the biomimetics community (see Nachtigall 2002, chapters 6, 10, 11; the BioKoN network working group). The case studies on ICT and nanotechnology (more precisely: nanobiotechnology) represent those technology developments that in fact focus on the model from nature but often do not formulate an explicit reference to biomimetics and bionics.[11] With the case studies on tissue engineering and – to an even higher degree – the converging technologies, the focus is widened once more. In the long term, it seems to make less and less sense to treat biomimetics as an isolatable scientific and technological field of development. If we expand our long-term view to ten years or more, the development of biomimetics must increasingly be considered in close context with the development of biotechnology and nanotechnology, and, looking even further ahead, together with the remaining converging technologies (ICT).

...and developing fields of biomimetic research and development

In the case studies, we focused on the following questions:

Central questions of the case studies

- What is the current dynamic in the field, and what are the central (and most promising) paradigms, questions and problem-solving approaches?
- What is the present position of biomimetics in the respective science and technology field?
- What are the technological challenges in the respective fields and what significance do biomimetic approaches hold for coping with these challenges?

The results for each case study are subsequently presented followed by a concluding summary as well as overall conclusions for the field of biomimetics in general.

Sensorics

Hardly explicit but many implicit references to biomimetics

Sensorics has ranked highly in biomimetics literature almost since its beginning in the early 1970s; however, in the survey and technical literature on sensorics we find no direct references to biologically inspired, biomimetic, or bionic approaches. Sensorics, now as before, is viewed mainly as a classic science and technology field of research and applied science and

11 Still, these research approaches play an important role within biomimetic research; robotics and sensorics, for example, are dependent on IT, and that the lotus effect, after all, can only be achieved by means of nanotechnology (see Cerman et al. 2006). Altogether, expectations are frequently expressed, that with the help of nanotechnology natural materials can be replicated in lighter versions (Popper et al. 1998). However in the relevant biomimetics literature these areas are scarcely dealt with. Neither Nachtigall (2002), nor Kesel (2006), nor Cerman (2006) give nanotechnology or IT their own section.

is also taught as such.[12] Even so, one frequently finds indirect and implicit references to biomimetics.[13] As an example, the analogy "sense organs – sensor technology" (i.e. sensors as technical noses, eyes, ears) is frequently found (Toko 2000).

Research and development on biomimetic sensors

Those few publications that deal explicitly with biomimetic sensors and sensor technology cover topics that have been addressed for many years and therefore can also increasingly be found in the biomimetics literature:

- compound eyes of insects (and the associated signal-processing and signal-controlling apparatus) and their technical implementation[14]
- technical smelling and tasting sensors and "sniffers" based on biological models[15]
- infrared detectors in beetles and snakes and their technical implementation (Bleckmann et al. 2004, Gorbunov et al. 2002, among others)
- electropositioning in weakly electric fish (Bleckmann et al. 2004, MacIver/Nelson 2001, among others)
- sensor-actuator systems in crustaceans and their application in robotics (Ayers 2004, McGruer et al. 2002)
- distance and speed sensors based on the model of neuromuscular spindles (Jaax/Hannaford 2004)
- sonar-based sensors for mobile navigation based on the model of the bat (Carmena/Hallam 2004a, Carmena/Hallam 2004b, among others)
- strain sensors based on insect models (sensilla) (Skordos et al. 2002).

The spectrum is relatively small in light of the immense number of existing sensors and sensor systems and the respective parameters to be measured.[16]

Sensorics terms close to biomimetics

Among the terms in the general technical literature on sensorics that can be understood in the sense of biomimetics and bionics are those that in part have already been mentioned several times, such as *intelligent,*[17] *smart* (Liu et al. 2005), *learning* (König et al. 2001, among others), *self-calibrating, self-monitoring, self-optimizing, self-testing,*[18] and *self-sustain-*

12 See, for example: Wilson (2005); Tränkler/Obermeier (1998); Gründler (2004); Bernhard (2004); Jackson (2004); Furukawa (2004); Stibor (2005).

13 For example, Jackson (2004, 205): "However, artificial neural networks do not use logic. They require no prior knowledge or skills from an expert to be programmed into the software. They are systems that attempt to mimic the learning processes of the brain."

14 See also Viollet/Franceschini (2005), Ruffier/Franceschini (2005), Toko/Habara (2005), Argyros et al. (2004), Higgins/Pant (2004), Franceschini (2004), Webb et al. (2004), Schenato et al. (2004), among others.

15 See also Settles (2005), Toko/Habara (2005), Webb et al. (2004), Winquist et al. (2004), Li et al. (2001), among others.

16 See the annual surveys of developments and suppliers in the sensorics marketplace (in Germany) by Schanz (2003, 2004, 2005, 2006).

17 See also Liu et al. (2005), Meier/Smajic (2005), Mackensen/Kuntz (2004), König et al. (2001).

18 See also Gausemeier et al. (2005), Liu et al. (2005), Kanoun/Tränkler (2004), Prock (2004).

ing (Mackensen/Kuntz 2004, among others). What these functionalities have in common is that up to now they have only been utilized in natural systems, i.e. in organisms (biological systems) first. Terms such as *intelligent, smart,* or *able to learn* were not ascribed to technical systems such as machines, nor an autarkic *'self'* with the capability of acting alone. The use of these terms as attributes for sensors and sensor technology demonstrates a more or less considered transfer of natural/biological characteristics to these sensors.

Intelligence = biomimetics?

Yet, one should be cautious about the validity of such attributions. Often the "intelligence" of a sensor system is limited to the fact that the sensor can communicate with other sensors in a decentralized network,[19] which indeed might be a useful technical innovation, but has little to do with intelligence in the biological sense. Often, it is unclear what exactly is meant by "intelligence".[20] *In most cases*, the attributes of intelligence refer to the manner of signal processing;[21] the same is true with the use of the attribute "smartness."

The prefix »self«

Sensor (system) attributes with the prefix "self-" are more clearly based on a biomimetic approach, even if it is not explicit. The idea of acting on one's own initiative or independently definitely has a natural or biological origin. Technical systems have always depended on the operation or handling and influence of humans (from the first instrument up to the computer). Self-calibrating, self-diagnostic and self-optimizing systems cast off their dependence on human control and begin to resemble "autonomous" biological systems (organisms).[22]

Challenges

Challenges of sensorics

Among the most significant challenges and trends in sensorics development today are, first and foremost, increasing efficiency (Kenny 2005) with respect to sensitivity (input/output signal ratio), maximum detection limit (most extreme measurable value), greater resolution (smallest measurable unit), range of operation (with regard to measurements/phenomena), greater selectivity (for substances to be measured, above all, with biosensors and biochemical sensors), maximum linearity (non-linearity = percentage deviation from ideal calibration curve), low noise levels (background noise), shorter response times (time duration between input and output signals),

19 See, for example, Mackensen/Kuntz (2004).

20 See, for example, Meier/Smajic (2005).

21 See, for example, König et al. (2001).

22 "We use the term 'self-optimisation of a technical system' to mean the endogenous adaptation of the system's objectives to changing environmental conditions and the resultant autonomous purposive adaptation of its parameters, possibly also its structure, and thus its behaviour. Self-optimisation thus goes considerably beyond the familiar rule-based and adaptive strategies; self-optimisation facilitates systems with inherent 'intelligence' that are able to take action and react autonomously and flexibly to changing operating conditions." (Gausemeier et al.2005, 316)

and minimal hysteresis (measurement deviation due to cyclical variations in the parameter being measured). An improvement to sensor performance leads to greater efficiency in the overall (technical) systems in most applications where sensors are used.[23] As an example, more precise sensors for measuring parameters such as temperature, pH, concentration, pressure, etc., in a chemical process can contribute to a more optimal process run and greater efficiency, thus saving material, energy and time.

Requirements of future sensor developments

On top of the general operational requirements for their application, sensors and sensor systems have to meet further requirements. Future developments are intended to make them cheaper, capable of functioning in new, extreme environments, that are easier to handle as well as and capable of integration in other sensor, information processing, and actor systems; as well as greater mobility in their application. Obviously, not all of these requirements can be met simultaneously (see Abele et al. 2005, Engin et al. 2005, Rodriguez-Mozaz et al. 2005, Andreescu/Omowunmi 2004, Kanoun/Tränkler 2004, among others).

With respect to biomimetics, the demands being placed on smell (Settles 2005) and taste sensors (Toko/Habara 2005, Toko 2000) are especially noteworthy. Scent and taste are based on "complex, multi-component compound substances" (Gründler 2004); i.e. a large number of different substances must be recorded and simultaneously processed with regard to possible effects. This means not only large demands for the receptor elements and the transductor, but especially for the subsequent signal processing.

Major trends in sensorics

Among the important trends in sensorics are: a) miniaturisation of sensors and the subsequent signal processing,[24] b) integration of sensors and signal processing,[25] c) wireless sensor networks,[26] d) self-calibrating/monitoring/optimizing sensors,[27] e) multi-sensor systems,[28] f) gene technology for receptor elements and biosensors (Rodriguez-Mozaz et al. 2005, Tränkler 2000), g) molecularly imprinted polymers,[29] and h) online and inline sensorics (Abele et al. 2005).

Although this compilation is not complete, it offers an overview of the currently most important developments. In particular, the current miniaturisation of sensors and sensor (controller-actuator) systems and the development of wireless sensor networks are clearly dominating development in

23 See earlier for the numerous areas of application of sensorics.

24 Micro-/nanoelectronic mechanical systems (MEMS/NEMS), see Engin et al. (2005), Rodriguez-Mozaz et al. (2005), Gründler (2004), Jackson (2004), Kanoun/Tränkler (2004), Mackensen/Kuntz (2004), Schanz, (2004).

25 Single-chip sensor systems, lab-on-a-chip, see Engin et al. (2005), Jackson (2004), Mackensen/Kuntz (2004).

26 Wireless sensor networks, see Engin et al. (2005), Puccinelli/Haenggi (2005), Rodriguez-Mozaz et al. (2005), Kanoun/Tränkler (2004), Mackensen/Kuntz (2004).

27 See Gausemeier et al. (2005), Gründler (2004), Kanoun/Tränkler (2004), Prock (2004).

28 See Rodriguez-Mozaz et al. (2005), Gründler (2004), Kanoun/Tränkler (2004).

29 MIPs for bio(chemical) sensors, see Rodriguez-Mozaz et al. 2005

almost all fields of sensorics. Here, sensor technology is obviously following the latest technology developments in other fields. This also applies to the increasing multi-functionality of the devices, which can be seen in the rapid development of multi-sensor systems. More sensor-specific, however, is the tendency to increasing development of self-calibrating and self-monitoring sensors with the objective of better handling and system reliability.

The fulfillment of some requirements will lead to conflicts with others; at the same time, some of the trends can be understood as answers to the challenges presented above. For example, wireless sensor networks thus permit greater mobility and application in previously inaccessible environments. Miniaturisation may likewise contribute to lower energy consumption (Puccinelli/Haenggi 2005) and open up new fields of application, e.g. in minimally invasive and non-invasive medicine (wearable sensors) (Engin et al. 2005, Andreescu/Omowunmi 2004).

Robotics

Robotics: traditionally close references to bionics

A biomimetic approach is already inherent in almost all characteristics of robotics. Mankind was already the model when Karel Çapek coined the term "Robot" (worker) to name the human-like machine in his 1921 stage play (Ichbiah 2005, 40). Moreover, robots (regardless of terminology) always have been constructed for the purpose of more or less intentionally emulating – or even exceeding – the functionality of humans and later animals. The typical stationary industrial robot emulates the movements of the hand and arm of the former varnisher, welder, or mechanic for example; a function partly made possible by applying the principle of the joint. Even though many mobile robots still move on wheels – a form of locomotion having no natural counterpart – their development has always been closely connected with the objective of creating systems capable of fulfilling several characteristic functions of animate organisms, especially independent locomotion in space, perception of the environment and response to environmental stimuli as well as the (situation-dependent) objective-oriented execution of certain activities. Such capabilities for "behavior" in a biological sense are achieved in (mobile) robots by the combined use of coupled sensors, (intelligent) controllers and actuators (Selverston 2002). Furthermore the development of "intelligent" controls has always been a focus of robotics,[30] simultaneously underlining the striving for (artificial) machines with natural attributes or resembling nature.

Many explicit references to biomimetics

In addition to the generally implicit reference to biomimetics in robotics, there are numerous biomimetic references in individual robotics developments (see, e.g. IEEE 2004; Dario et al. 2005; and the survey of

30 Although the term "intelligence," as mentioned already, is often not clearly defined. As a rule, however, what is meant is a capability that distinguishes (higher-level) organisms and is alien to technical systems.

Bar-Cohen/Breazeal 2003). The space agencies in particular are interested in developing robots having autonomous qualities (e.g., ESA 2004); they refer explicitly to autonomy (ESA 2004) and the self-organizing capabilities (Pettazzi et al. 2005) of natural systems, which should be emulated. Finally, there are a few developments with the goal of "reconstructing" a complete higher-level organism, e.g. the artificial rat project (Meyer et al. 2005; Filliat et al. 2004). Admittedly these are distant goals where – in most cases, only some components of the developed robots are actually being designed and implemented in accordance with a biological model.[31]

Challenges

Robotics development can be broken down into three components: a) actorics (drive controls – or actuator technology) – the fixtures a robot actively uses for locomotion in its environment or for manipulation; b) sensorics – devices for collecting information on the environment; c) controls – the linking element between actorics and sensorics. Key challenges in robotics can be identified in all three areas.

Actuating elements

Challenges in the area of actorics: The natural models are superior in their maneuverability, functional range, flow characteristics (fish and birds), and energy efficiency. The use of robots currently is problematic or impossible in applications in which non-rigid or flexible parts must be handled, rapid and low-loss transmission of force is necessary, the energy-power ratio must be kept very low, or for safety reasons close contact with people is necessary.

Challenges in the area of sensorics: Biological systems for sensory perception and signal processing are superior to today's man-made systems. It is a matter of sorting the information by relevance and reducing the amount of data processing necessary. The technical emulation of touch, smell and taste also represents tremendous challenges.

Control

Challenges in the area of control: Traditional control mechanisms are not sufficient for the tasks a robot is to carry out. Therefore scientists are trying to emulate the control principles of the brain, especially decentralised control and parallel computing. Realisation of robots having autonomous characteristics (ESA 2004) requires coping with the challenges in all three areas.

31 Apart from this, it is questionable, whether such a "replica" still falls under our definition of biomimetics, and whether such a project is even worthwhile.

The material sciences

Blurring of basic and applied research

The material sciences are a field of research and development, in which the borders between fundamental research and applied research were eliminated right from the very beginning. The foundation of the material sciences is in the engineering disciplines; for a long time, its development was governed on the scientific side by solid-state physics, physical chemistry, quantum mechanics, and, last but not least, mathematics. To this was added metallurgy, polymer chemistry, and ceramics, with their mixture of scientific and (engineering) practical knowledge. Since the early 1990s, the history of the material sciences has been characterized by the heavy influence of the computer sciences (the computational material sciences), nanotechnology and biology. As with biomimetics, the material sciences, can also provide a venue for the converging sciences and technologies.

The French science historian Bernadette Bensaude-Vincent (2001) has traced the development of the material sciences in an interesting research project from its beginnings in the 1950s until today. According to her findings, the biomimetic idea of learning from nature entered the material sciences in the last decade of the 20th century and in parallel as a part of nano-technology and computational material science. Rightly, she points out that Feynman, in his famous talk "There's Plenty of Room at the Bottom," referred to biological models and also notes that elements of biomimetics have been introduced via computational material science. After all, the simulation of evolutionary processes plays an important role, for example, in combinational chemistry. Ultimately, it was the fascination with the capabilities of biological materials (anisotropic, hierarchically structured, intelligent, smart, adaptive, self-healing) that determined the focus on natural models. The opportunity of learning to deal with this across-scale complexity in an engineering technical manner was of special importance. Certainly Stephen Mann (Mann et al 1989, Mann 1996) must be included among the pioneers of this capability in the form of template-controlled crystallisation, which had numerous precursors in colloidal chemistry and molecular biology.

Orientation towards natural model

»Learning from nature« very common

An analysis of current review articles[32] confirms that the idea of learning from nature is widespread within research in the material sciences, although not all approaches inspired by nature admit being biomimetic.

Nanotechnologies and self-organisation

In some biochemical and biotechnological approaches, the biomimetics references are only visible at second glance. Scientists who develop "smart" and biocompatible materials seldom mention their focus on the model of nature, but the functionalities that the researchers intend to implement have already been realized in animate nature. The research process hardly differs in the analysis from the mainstream in chemistry and bio(techno)

32 See Langer and Tirrell (2004), Anderson et al. (2004), Sanchez et al. (2005).

logy: Scientists investigate molecular structures, try to understand the functionality of specific compounds, etc. The synthesis approach, however, follows the principles of molecular self-organisation and therefore is a biomimetic approach. Thanks to diverse developments in biology and chemistry as well as in the material sciences and nanotechnology, new opportunities are opening up to better understanding self-organizing and self-structuring properties of molecules and to make use of these for new applications.

Other approaches mentioned in the review articles are biomimetic: in these cases, the objective of the researchers was to emulate such properties of natural surfaces or interfaces as hydrophilic, hydrophobic or adhesion (Sanchez et al. 2005) and to convert them into technical products.[33]

Challenges

Trends in materials sciences

An example for the classification of future trends in the material sciences can be found in Table 5.[34] The Japanese National Institute for Material Science (NIMS), categorizes the material sciences into two levels, the first level – in the sense of a differentiation between *technology push* and *demand pull* – distinguished between: a) "characteristic seed materials" (among these, nanomaterials, superconductive and magnetic materials, and semiconductors) and b) "needs-oriented materials" (among these biomaterials, ecological materials and high-temperature materials for turbines and

Table 5 — Classification of the material sciences

Characteristic seed materials	Needs-oriented materials
(1) Nanomaterials	(5) Biomaterials
(2) Superconducting materials	(6) Ecomaterials
(3) Magnetic materials	(7) High-temperature materials for jet engines and gas turbines
(4) Semiconductor materials	

Performance-seeking materials	R&D materials
(8) Metals	(12) Analytical and evaluation techniques
(9) Ceramics	(13) High-field generation techniques and applications
(10) Composites	(14) Nano-simulation science
(11) Polymers	(15) New material creation techniques
	(16) Acquisition and launching of materials data
	(17) International standards

[Source: NIMS 2005, 5]

33 Examples mentioned: Bico et al. (1999), Sanchez (2001), Shin et al. (2003).

34 See Foresight (2000) for an alternative.

the like). On a second level, they differentiate between c) "performance-seeking materials" (high-performance materials, such as metals, ceramics, composite materials and polymers), and, finally, d) "R&D materials" (material science measuring, simulation, and development approaches).

This Japanese study, like the British foresight study, assumes technological challenges across the entire spectrum of the material sciences (NIMS 2005; Foresight 2000). They see only a marginal role for biological and biomimetics approaches. They are only mentioned in the NIMS study in the areas of biomaterials (superbiocompatibility, induction of cell growth and cell differentiation, nanobiotic materials, bioelectronics, lab-on-a-chip) and ecomaterials.[35]

The rapidly emerging field of biomimetic materials will form one of the most important technologies of the 21st century

However, the authors of the white paper on the material sciences at the Max-Planck-Society write: "The rapidly emerging field of biomimetic materials will form one of the most important technologies of the 21st century" (Rühle et al. 2001, 12). Bensaude-Vincent – who agrees with the significance of this trend (2001) – explains the relevance of biomimetic materials with the fact that a focus on the model of nature opens new approaches in research (Bensaude-Vincent 2001). Products would no longer be manufactured top-down, i.e. cut out of a block in the way that a sculpture is chiseled out of a rock, or by shaping, she suggests; nor would material be ground or liquefied and then poured into a mold. Instead, materials and products could now be manufactured "bottom-up," i.e. through the alignment of individual atoms or molecules, for example, by means of their own self-organisation. Only in this manner does it seem possible to create such varied hierarchically structured materials as bones, teeth, mother-of-pearl, and spider-silk. However there are even more challenges, these involve adaptedness and adaptability to the environment, i.e. adaptive and self-healing materials. The arguments put forward by Rühle et al. (2001) likewise continue on from Bensaude-Vincent's line of reasoning:

Bottom-up production processes

> *"'Mother Nature' uses highly smart molecules to create her objects by relatively simple and robust processing routes, whereas traditional materials science usually applies sophisticated processing routes to create artificial materials from simple elements" (ibid., 122).*
>
> *"We see the future of materials technology as the combination of two developments: The convergence of top-down and bottom-up processing to overcome the 30 nm barrier. The connection of traditional semiconductor materials (Si, SiO2, etc.) with organic/biological materials, possibly with 'small-scale structured' interfaces. This path would open up plenty of new technological fields" (ibid.).*

35 Purification materials with sophisticated sensing functions need to be developed by using nanotechnology, self-organisation, and template reaction, p. 7.

Figure 1: The molecular bioengineering fields

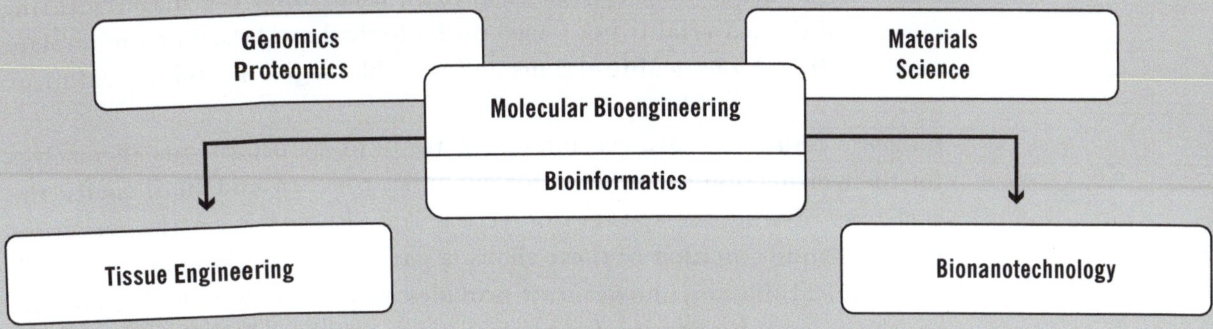

[Source: Rühle et al. 2001, 123]

> *"More generally, biomimetic materials processing is leading to a new interdisciplinary forum which could be called molecular bioengineering. We see three major roots for this new discipline: (i) molecular cell biology with contributions from genomics and proteomics, (ii) materials science, and (iii) bioinformatics/biomodelling" (ibid. 123).*[36]

As the most important fields of molecular bioengineering, they suggest: new biomaterials, tissue engineering, bionanotechnology, biosensors, and biomimicking (ibid. 125)

Technological challenges of materials sciences

Among the greatest technological challenges in the material sciences today, are those of medical technology and prosthetics. The task here is to create artificial materials that can substitute for teeth and bones, but also soft materials, such as heart valves. The same applies to biocompatible joining technology. Bonding techniques and adhesives are the special challenge here.[37]

Bioceramics

Separation of biomimetic approaches

In the field of bioceramics, it becomes especially clear, which difficulties exist in making a clear-cut distinction between biomimetic and non-biomimetic approaches within the material sciences. Basically the field of bioceramics can be subdivided into the following branches, which can be classified more or less clearly as biomimetic approaches depending on the viewpoint and interpretation of biomimetics (see Grathwohl 2004 as well as Figure 2):

36 Against this backdrop, the Max Planck Society makes the following recommendations: "The European materials science community should be allied with leading institutes in molecular cell biology and bioinformatics to establish joint European Research Centres in Molecular Bioengineering supported by national and EC funds... Moreover research platforms in important areas of materials science should be initiated. Possible themes include: 'Small-scale materials science and engineering,' 'bioengineering of bone and cartilage,' 'materials processing on biomolecular templates,' 'cellular machines and bioengineering" (2001, 124).

37 "It is not yet possible to get durable bonds between artificial organs and human tissue" (Rühle et al. 2001, 70).

Biofunctional ceramics – in endoprosthetics, e.g., implants of or in teeth and bones. This is the preferred term to be used when the manufacturing process of the material is not based on biological models, but the implant merely fulfills biological functions and possibly is given a biological form.

Biogenic ceramics – are ceramics that are built by organisms themselves for their protection or support, for example, conches and snail shells, the skeletons of siliceous sponges or vertebrates. Research on the structure, function and formation of these shells is part of the fundamental research in (technical) biology; however, it provides an extremely valuable (and often the crucial) foundation for biomimetic research and development.

Biomorphic ceramics – are ceramics during whose manufacture biological structures have a formative influence. As an example, in a current line of research development, wood is initially carbonized; the cellular structure of the wood is maintained to a great deal in the resulting carbon structures. These carbonized structures are then soaked with silicic acid, causing the latter to bond with the carbon (silicon carbonate ceramics). The question whether this procedure can be called a biomimetic one or not is not easily answered.

Biomimetic ceramics – are ceramics whose manufacture is based on biological models. The template-controlled manufacture of artificial nacre (mother-of-pearl) for various applications in surface technology, which is presently under investigation at the University of Bremen in several projects, falls into this category; this project is clearly a biomimetic one (see Fritz et al. 2005).

Bioceres – are composites consisting of a biological and an inorganic component. The biological component can range from the molecular level (e.g. enzymes or DNA) to living bacteria. If the inorganic component has nothing to do with biological models, it is biotechnology rather than biomimetics, as defined here. If dealing with the the bond between a computer chip and a nerve cell, differentiation would be more difficult, because this would be heading in the direction of converging technologies.

Challenges

In the technical literature,[38] the following main trends are described: artificial bones and teeth, development of fracture-resistant ceramics, and emulation of optical properties.

38 Special section "Materials and Biology" of SCIENCE from November 2005 (vol. 310) as well as January 2006 (vol. 311).

Figure 2 — Structure of the field of bioceramics

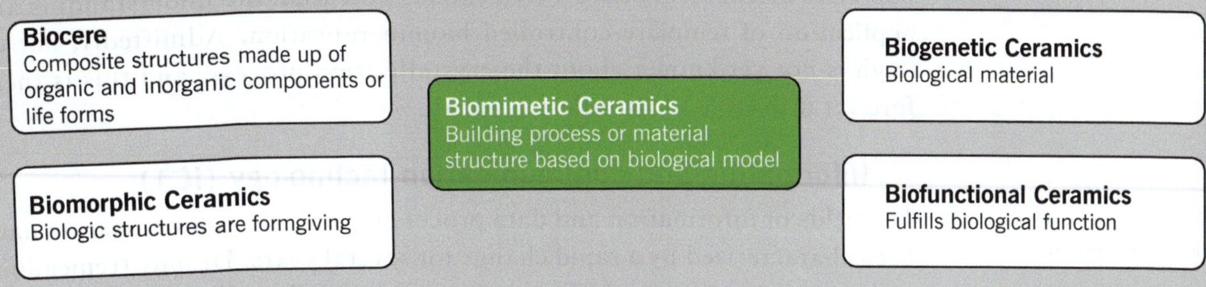

[Source: authors' own representation following Grathwohl 2004]

Artificial bones and teeth

The development of artificial bones and teeth involves the manufacture of absorbable and cell growth–inducing ceramics. The focus of research is on the emulation of hierarchical levels down to the nanometer realm (hierarchical structuring) and the interaction with cells. Only with the hierarchical structuring of not-yet-realizable combinations of ceramic properties, such as fracture resistance and fracture toughness, such techniques become possible. With the latter, the three-dimensional structures on the different levels, i.e. in the meso-, micro-, and nanometer realm, as well as the surface properties, play a large role. They serve as a strong mechanism to promote and regulate cell growth (Stevens and George 2005).

Sponges as model

Above all, conches and sponges, with their outstanding mechanical properties, serve as a biological model for the development of fracture-tough ceramics. Here the goal is also the emulation of the hierarchical levels in order to realize the desired mechanic properties (Mayer 2005).

Some of the optical properties of natural ceramics most desirable for emulation are, the special forms of light conduction and collection found, for example, in sponges and starfish. Important factors are the material combinations as well as the structural compartmentalisation (Weiss 2003; Lee and Szema, 2005).

Importance of visualising procedures

Among the developments that could give the field of bioceramics an impetus are new image-producing methods (image-producing procedures such as low-angle x-ray diffraction, infrared spectroscopy imaging, and microtomography). They make new insights into the structures and properties of such natural nanocomposites as nacre, collagen, and wood cell walls, and in part, the mineral-collagen composite in bones possible.

Secondly, the use of methods from biochemistry makes it now possible to understand how molecules are formed and react; this allows us, among other things, to pursue biomineralisation.

Self-structuring procedures

Finally, there is great potential in the development of self-structuring process methods. In most organisms, biomineralisation occurs by means of an amorphous precursor with special properties. Investigation of this operative mechanism offers a tremendous potential: the understanding and application of template-controlled biomineralisation. Admittedly, not enough is not yet known about the crystallisation process, and this is therefore yet utilizable to its fullest capacity.

Information and communication technology (ICT)

High dynamics of information and communication technologies (ICT)

The fields of information and data processing and telecommunication have been characterized by a rapid change for several years. Despite tremendous wide-ranging successes, ICT scientists face a number of challenges. Above all, they must deal with problems of capacity, increasing complexity, and the ongoing miniaturisation of ICT systems.

Traditional problem solving vs. new complexities

It is in these areas, where traditional solutions reach their boundaries, where alternative problem-solving approaches are beeing sought. Promising alternatives are seen in quantum mechanics (e.g. quantum computers), in optics (e.g. optical fiber cable), and also in biology (e.g. evolutionary techniques, DNA computers, and emulation of the brain and the nervous system). According to Sörensen (2004), the interest in biologically-inspired solutions applies in a similar way to the general problem of complexity, as in the case, according to Bensaude-Vincent, in the material sciences: "The general popularity of biology is a consequence of a general scientific movement towards theories of systems, complexity, and processes. This development not only describes the historical genesis of biomimetics but also – so I will claim – its raison d'être" (ibid., 211). The idea of using nature as a source for innovations and solutions in the field of information science is already found in the work of Turing in the 1940s (see Paun 2004).

Terms such as bionics and biomimetics are no longer used...

... rather »evolutionary algorithms«, »evolvable hardware«, »organic computing«

Numerous research projects at many different levels in ICT do not use the term *biomimetics* nor *bionics*; instead terms such as *bio-inspired* or *evolutionary* are used (e.g. "evolutionary algorithms," "evolvable hardware," and "bio-inspired computing") (Paun 2004, de Casto 2005). Examples can be found in the IBM approach "autonomic computing"[39] and the DFG-supported research emphasis "organic computing."[40] Both refer explicitly to the vegetative nervous system (Kephart and Chess 2003) and the behavior of living creatures (Müller-Schloer et al. 2004a) and utilize genetic algorithms (Trumler 2006), but neither uses the terms *biomimetic* or *bionic*.

Sörensen reflects on the difference between *bio-inspired* and *biomimetic* in his highly regarded dissertation on "Ambient Ecologies – Toward Biomimetic IT" and, in addition to the already mentioned complexity, underlines

39 See Kephart and Chess (2003): http://www-03.ibm.com/autonomic/about.html

40 German National Science Foundation (DFG Schwerpunktprogramm) 1183, est. 2004: http://www.organic-computing.de/SPP

functionalities such as self-organisation and adaptivity as being constitutive of the biomimetics approach (Sörensen 2003, 76; Sörensen 2004):

> *"So the difference between merely 'inspired' approaches and biomimetics is that biomimetic models relate to models of complex adaptive systems in non-trivial ways. Biomimetic models must capture essential traits of self-organization and adaptivity by investigating complex adaptive systems so that the same general laws of self-organized adaptivity will govern both technological and natural complex systems."*
>
> (SÖRENSEN 2003, 76)

Indirect reference to biomimetics

It becomes clear that in the field of ICT as in other fields, the terms biomimetics and bionics are rarely used. Within the ICT field, however, many sub-fields have arisen in which the use of natural systems as a model for the development and/or optimisation of technical systems can clearly be seen. In many cases, the biomimetic nature of these approaches is already clear in the terminology used to describe the approaches: "natural computing," "evolvable hardware," etc.

Challenges

Ubiquitous computing

Rapid increase of data volumes being processed is one of the major technical challenges that the field has faced for some time now – as well as the ever-growing computerisation of our surroundings (ubiquitous computing).

Limits of data processing and miniaturisation

At the same time, the physical limits of traditional approaches have become apparent. This especially holds true for the ongoing miniaturisation, in which the materials in use are reaching their operational limits (e.g. silicon in computer chip production) (BSI 2003): firstly the speed in which a processor performs arithmetical operations will not exceed 5 GHz due to the laws of thermodynamics (Microsoft 2005, 14); secondly, it is not possible to arbitrarily increase the speed at which data is exchanged between processor and data storage medium (ibid.); third, the miniaturisation of transistors on the computer chip will reach its limits when transistor size begins to approach that of single atoms and quantum mechanics effects begin to dominate (if not sooner), leading to unpredictable behavior; this limit may already be reached in 10–15 years time (ibid., Fulk 2002).

At the same time, computerisation and miniaturisation increase the complexity of these tasks and the operative systems; achieving this while maintaining functionality will, in addition to the other tasks, be one of the greatest challenges in the field of computer sciences and information technology. The prospects for success by means of traditional approaches are rather limited at present. One significant hurdle for researchers may be the "machine centricity" of the current IT infrastructure. The „Organic Computing" initiative speaks of "algorithmic sharing of work" between man and machine in this respect: the human designs algorithms and thus is responsible for the creative aspect, and the machine executes the given program step by step.[41] The disadvantage in this distribution of tasks is that the human must be able to understand each step of the machine to be able to find mistakes in case of a malfunction. This top-down approach requires the researcher to have an overview of the entire structure from which to work and then solve an individual problem or make an improvement within the framework. It is clear that this procedure may work with those algorithms where the structure can easily be followed, but the more complex the IT application becomes, i.e. the longer the algorithms become, the harder the task for the programmer to keep track of the individual steps.

The limitations of traditional approaches also have an economic perspective: with the growing complexity of IT applications and their use in such safety-relevant fields as the automobile industry, medicine, and aviation, development becomes increasingly more time-consuming and costly. Companies must employ more personnel, drastically increasing development costs.[42] At a certain point this practice will reach its economic limits of feasibility. (Müller-Schloer et al. 2004b).

Self-x-attributes

See chapter 6.3

See chapter 6.4

A promising trend for coping with these challenges is the development of IT systems having *self-x properties* (Kephart and Chess 2003, Kasinger 2005):

- self-configuring
- self-optimizing
- self-repairing
- self-protecting

Self-X approaches may be considered to be biomimetic because they involve solving complexity problems by emulating the properties of natural systems.

Prosthetics

Prosthetics in wide biomimetics or bionics concept

Within the field of bionics, developments in the area of prosthetics play an important role – and, in turn, within the field of prosthetics, bionics likewise do so. Nachtigall writes that anthropotechnology, the adaptation of machines to humans and their needs, is one of the "most prominent research tasks in the field of extended bionics" (Nachtigall 2002, 287). He thus establishes very broad references to bionics and names as an example the development of automobiles of the future that represent a "biomechan-

41 See also: www.organic-computing.org

42 According to IBM, in businesses today 70% of the IT budget is paid out for labor (Enterprise Management Associates 2006, 1).

ically adapted optimal shell" for the "biomechanic system man" (Nachtigall 2002, 288).[43]

Bionics in anglo-saxon world sometimes synonymous with prosthetics

The terms *biomimetics* and *bionics* are used extensively in the technical literature on prosthetics. Particularly in the English language realm, the term *bionics* is used, as a rule, *exclusively* refering to prosthetic systems in the broadest sense. As a representative of a multitude of examples using the terms, Herr et al. (2003) shall be mentioned here. Bionics represents here the understanding and technical application (and not necessarily a direct copy) of the functions of the human body.

From wooden cane as leg prosthesis to exact copies of human limbs and organs

The terms *biomimetics* and *bionics* are not used consistently within the field of prosthetics research, but their various uses can be fully integrated. Formerly, prostheses were exclusively inspired by the shape of, e. g., natural limbs. They were manufactured, so to speak, as a formal analogy to the original, but were later gradually improved as knowledge and scientific development advanced.

The wooden stump as leg prosthesis was more an analogy than a copy. With further scientific progress in the various fields of biology and medicine and the resulting opportunities for greater understanding, analogy gradually gave way to the goals of replication and imitation. At the level of practical development of prostheses, analogy and replication cannot always clearly be distinguished from one another, but it is obvious that the expanded knowledge increasingly brings reproduction closer to the realm of the possible. From this perspective, one could bring together both the emphasis on gradual improvements as well as the goal of biomimetic replication under the umbrella term *bionics*.

Challenges

The copy as objective of prosthetics

The central objective of prosthetics research is to come as close as possible to the natural functions of organs and limbs to the point that it becomes possible to fully reproduce them: "In the next several decades, continued advances in human-machine neural interfaces, muscle-like actuators, and biomimetic humanoid control schemes may result in dramatic improvements in the quality of life of the physically challenged" (Herr/Whiteley/Childress 2003). Three key concepts play a central role in the debate on the future of prosthetics: *neuronal control, biohybrid systems,* and *muscle-like actuators.*

To demonstrate current development trends, some examples of fundamental research are described in the following.

Example visual prosthetics

The area of visual prostheses involves efforts to restore sight while bypassing the retina and the optic nerve; to achieve this data from a camera is

43 In the 1993 VDI publication "Technologieanalyse Bionik: Analyse & Bewertung Zukünftiger Technologien," already, "bionic prosthetics" is assigned to the field of bionics with the following subcategories: "function optimization of prosthetics," "in situ prosthetics," and "improving tolerability" (VDI 1993, 15).

sent directly to a microprocessor implanted in the brain, which then stimulates the area of the brain responsible for vision, with the intent to generate visual impressions. The technology was developed by the Portuguese-based Dobelle Institute.[44] Blind persons who lost their sight after childhood have reported being able to distinguish between light and dark after implantation (Nsanze 2005).

Example artificial hippocampus

An artificial hippocampus is also presently under development; this consists of a computer chip that not only stimulates the brain, but also takes over the activities of a damaged hippocampus – essentially memory functions. Such research efforts in the field of neuronal control exceed by far those in the field of prosthetics and delve into the area of improving performance; however, researchers do not expect to have a chip ready for human implant for another ten years (Nsanze 2005).

Steering wheel chairs by thought

The company Cyberkinetics reports work on a chip called "Braingate," which is supposed to make possible the control of objects by means of thought. The chip system consists of a sensor, which is implanted into the motor cortex of the brain and an apparatus that interprets the brain signals. When certain actions are thought, the corresponding electrical impulses are not sent to the limbs but are instead recorded by the chip system and then transmitted, interpreted, translated, and sent to a computer, which can then be operated by means of human thought. The chip was designed for paraplegic persons and is now in clinical testing. In the future, it is intended to be used for the manipulation and control of wheel chairs and exoprostheses.[45]

Exo prosthetics

There are also efforts in the field of exoprosthetics[46] to combine new prostheses with cultured tissue underway. For instance, Hugh et al. developed small robots at the Massachusetts Institute of Technology that are driven by frog muscles (Rötzer 2001). The muscles are connected with the mechanical components and are activated by stimulation with artificial current pulses. It is the long-term objective of the scientists to exchange the presently used comparatively heavy electric motors in exoprostheses for explanted human muscle tissue (ibid.). In the respective publications, the scientists classify their research work as being in the field of biomechatronics, but the keyword index also lists the key word *bionics* (Herr/Dennis 2004). This line of development is carried on further under the term *biohybrid body parts*. Scientists hope to use cultured tissue, extended bone stumps, titanium prostheses, and a new generation of implanted sensors to create prostheses that can be controlled by signals from the brain (Versweyveld 2004).

44 Web: www.artificialvision.com/ (no longer available).

45 See www.cyberkineticsinc.com for a detailed description.

46 Prosthetics located outside the body.

Role of nanotechnologies in prosthetics

Nanotechnology will also play a major role in the development of future prostheses, for example, in the combination of exoprostheses with artificial tissue, in the biomanufacturing of implants (National Institutes of Health Bioengineering Consortium 2000), and in the development of artificial muscles for exoprostheses (Bar-Cohen 2001). The publications mentioned also make extensive references to bionics. The Health Bioengineering Consortium report only refers to cell-based drug carriers, but Bar-Cohen explicitly states: "Practitioners in biomimetics, a field where robotic mechanisms are developed based on biologically inspired models, are particularly excited about these materials" (Bar-Cohen 2001, 45).

Tissue engineering

Tissue engineering as marginal topic of German biomimetics…,

Tissue cultivation plays only a marginal role in the biomimetics community. It appears neither as a biomimetic field in the comprehensive work by Nachtigall (2002) nor in the technology analysis of the VDI (1993); it also is not explicitly mentioned in a recent report issued by the BMBF on biomimetics. Only in a subchapter on "Humans and material" are the principles of bone growth mentioned as a model for the development of hard tissue implants (BMBF 2005). The European Space Agency mentions tissue cultivation only a side topic its survey on biomimicry. In the chapter on processes, cell division processes are mentioned as exemplary activities for the bottom-up production of materials. Organismic production of tissue by cell division and differentiation is explored in the context of nanotechnology and the desire to artificially reproduce these processes.

The German Federal Ministry of Education and Research (BMBF) is presently promoting research that focuses on the subjects of tissue engineering and the biological replacement of organ functions under the general heading of "regenerative medicine," but without referring to biomimetics or bionics (BMBF 2005).

… but permanent component of anglo-saxon biomimetic research

In contrast, in the English-language realm, tissue engineering is considered to be a typical example of biomimetic research. This is the case, for example, at the National Institute of Dental and Craniofacial Research, in the U.S. (NIDCR 2005). Biomimetics is characterized as the science in which biology and engineering come together in an effort to discover the biological principles necessary for the manufacture of bio-inspired materials for tissue cultivation (NIDCR 2005, 1).

Biomimetic explantation of teeth and bones

Many businesses strongly relate tissue cultivation to biomimetics through their products. The company Dentigenix, which intends to culture entire teeth and components of dental enamel, refers to biomimetic methods as elements of their development process; another company, Biomimetic Pharmaceuticals, is planning to culture tissue to replace damaged bone.

Nanotechnologies and tissue engineering

Those authors who relate tissue cultivation to nanotechnology also make such references to biomimetics (Prokop 2001, Wen/Shi/Zhang 2005). When speaking of biomimetics in connection with nanotechnological methods for the cultivation of human body tissue, it is generally a reference to controlling biological molecular formation and assembly processes.

Nanobiotechnology

Nanobiotechnology: a current trend

The effort by researchers to combine nanotechnology and biotechnology is relatively new (see Petschow et al. 2005). The term *nanobiotechnology* first appeared in 2000 in the context of the U.S. National Nanotechnology Initiative. So far, there has been no consistent definition of nanobiotechnology; differentiations from other disciplines have been rather vague.[47]

Close connection of nanobiotechnology and biomimetics

Since nanobiotechnology is more a description of a scientific trend than a well-defined field (or fields) of research, it is presently not possible to make a statement about the significance that biomimetics holds. However, a clear connection between this trend and biomimetics approaches can be observed, for instance, some attempts to define nanobiotechnology do explicitly refer to biomimetics or bionics (e.g. Goodsell 2004). Nanobiotechnology thus involves a new paradigm for production, a focus on the "production methods of nature" – particularly the bottom-up processes of molecular self-organisation. With the decoding of the organismic mechanisms of development and organisation, natural processes can, at least in principle, be emulated and subsequently implemented as "more natural" technologies. Hampp/Noll (2003) write: "Ever more detailed insight into the molecular function of biological macromolecules allowed a new branch of bionics to emerge: nanobionics." Subsequently, Hampp (2005) essentially equates nanobiotechnology with bionics /biomimetics.

Challenges

Bridge between animate and inanimate nature

Nanobiotechnology builds a bridge between animate and inanimate nature and attempts to understand biological functional units at the molecular level, as well as produces nanoscale functional components in a controlled manner, taking into account technical materials, interfaces, and boundary layers.[48] Such a bridge can be achieved from two different directions: from nanotechnology to biotechnology ("Nano2Bio") and vice versa ("Bio2-Nano").

47 "We are now witnessing the emergence of another field in which the physical, chemical and biological sciences are converging. This field is nanotechnology. And the clearest sign that nanotechnology is emerging is that nobody can quite agree on what it is." Editorial "Why small matters." Nature Biotechnology 21, 1113 (2003). For further remarks on the difficulties with definitions of nanobiotechnology, see Petschow et al. (2005).

48 See the nanobiotechnology Web page of the BMBF: www.nanobio.de. Detailed accounts can be found, for example, in VDI (2002), VDI (2004), the publications of Seeman/Belcher (2002), Ball (2001), Goodsell (2004) and Jones (2004).

Nano2Bio

"Nano2Bio" refers to the use of nanotechnological approaches in the investigation of biological issues (VDI 2002). Thus nanobiotechnology could prove to be a necessary key technology for many (potential) technical applications, e.g., for providing bottom-up production processes (building up molecular electronic structures), nanoscale neuro-electronic interfaces, or functional hybrid systems and biological layers.

- Fields of application are (VDI 2002):
- Analytics and drug research (in vitro): e.g. genomics, proteomics
- Diagnostics: antibody detection, biosensors, toxicity investigations
- Drug transport and deposition
- Biocompatible materials and surfaces

Bio2Nano

"Bio2Nano" stands for bio(techno)logical materials and structural designs for the production of functional technical nanosystems that can be applied in various fields (VDI 2002). In the pharmaceutical field, as well as medical technology and biotechnology, miniaturisation is in great demand; it will probably only thrive and advance with the incorporation of experiences and development gained in the physics and chemistry-dominated areas of nanotechnology. Examples in this field would include biochip technology, in vivo early diagnosis and treatment methods, nanoparticle-bound gene therapy, and also tissue engineering.

Three levels of application can be identified for "Bio2Nano", which are almost all characterized by a close relationship to biomimetics approaches[49]: a) nanofabrication and nanostructuring on the basis of bio-based methods – the objective being the use of chemo-physical principles of self-organisation of molecular units for complex structures; b) technical use of functional biomolecules or hybrid systems – this involves the application of isolated functional biomolecules in technical systems or in combination with non-biological elements; c) formation or design of interfaces between biological and technical systems or materials.

In the end, it is the concept of bottom-up production processes, together with the possibilities of self-organisation[50] that prove to be the essential new elements and promises in nanobiotechnology.

Converging technologies

Converging technologies: ICT, nano- & biotechnologies, materials & cognitive sciences

The situation with the converging technologies is similar to nanobiotechnology. The term *converging technologies* became internationally known through the anthology "Converging Technologies for Improving Human

49 Nanobiotechnological applications in the area of medicine are not taken into account here.

50 Technical applications of self-organisation have also meanwhile been given BMBF support (http://www.bmbf.de/foerderungen/5150.php).

Performance" (Roco/Bainbridge 2002), the published papers of a conference held by the U.S. National Science Foundation and the Department of Commerce in December 2001. Convergence refers here to a tendency of cross-fertilization and mutual complementation and interchange of research results, discoveries, and applications in various disciplines and technologies such that a new aspect emerges – or even a new discipline or line of technology. Among the cross-fertilising disciplines are information and communication technology, nanotechnology and biotechnology, and the material and cognitive sciences. For instance, the development of new materials whose manufacture is made possible with the help of nanotechnology will in turn allow for the development of new kinds of integrated circuits. Biomimetic approaches are thus explicitly addressed, in particular, "emulation" of the cell (Roco/Bainbridge 2002, 12) and the perceptual apparatus (ibid. 12, 18).

Reasons for convergence

Two reasons for the trend towards convergence of nanotechnology, biotechnology, ICT, and the cognitive sciences are mentioned in the literature: first, the capacity for analysis and design at the molecular or nanoscale level, second, the complexity of problems and specifications. "At the nanolevel, genes, bits, neurons and atoms all started looking like the same thing," wrote the authors of a report for the National Research Council of Canada (2003, 10). Therefore, the scientists who had until then done their own research on atoms, neurons, bits, and genes began more frequently to exchange information. The other reason is found in the complexity of the problems that scientists are attempting to solve (Priami 2005, 2). Until now, researchers have tried to cope with complex situations by breaking them down into their components in order to try to understand their interrelationship. Priami calls this approach "reductionist." He suggests that current challenges no longer permit such reductionism and calls instead for a "systemic" view of phenomena. "The basic idea is to view the systems as something that has its own behavior not obtained simply by gluing [together] the behavior of the systems' components of which we already have all the information." (Priami 2005, 2)

Merging of disciplines

The authors of most R&D foresight studies[51] agree that a change in the traditional way of research within the disciplines of ICT, the cognitive sciences, the material sciences, nanotechnology, and biotechnology is taking place. So far, individual disciplines have viewed themselves as independent entities and defined themselves through their respective tasks; research and education were also separate, and each discipline tried to pursue its own scientific objectives. This practice is changing, scientific success in one discipline enables discoveries and new applications in others. Researchers are therefore becoming dependent on one another and must learn to work together: the boundaries between the separate disciplines are blurring.

51 Roco and Bainbridge (2001), Nordmann (2004), Roco (2005), Lieshout et al. (2005), Silberglitt et al. (2006).

New opportunities related to convergence

The authors of technology foresight studies on converging technologies see new opportunities in this convergence. They hope that by means of these converging research and development programs applications can be realized that would never have been possible on the basis of individual approaches: intelligent, self-organizing, biocompatible, environmentally friendly, adaptable technical systems (e.g., adaptable genetically modified grains, ubiquitous information access, intelligent textiles, and much more; see Silberglitt et al. 2006, 17ff.). Roco/Bainbridge (2001, 18–19) even go so far as to list brain-to-brain interaction, sleep resistance, increased cognitive capabilities, curing mental diseases, and others as visionary ideas for the converging technologies.

Biomimetics as learning platform for converging technologies

Biomimetics, the material sciences, and nanotechnology anticipate to a certain extent these coming developments in that they already require in their very nature the exceptionally close cooperation of biology, chemistry, physics and engineering. The current trends in biomimetics, the material sciences, and nanotechnology thus can also be understood as an important learning venue, where valuable experience can be acquired and suitable approaches and concepts for dealing with the converging technologies developed.

Roco (2005), going even further, assumes that the current convergence of technologies we are seeing is a temporary process that will reach its climax

Figure 3 — Development of converging technologies

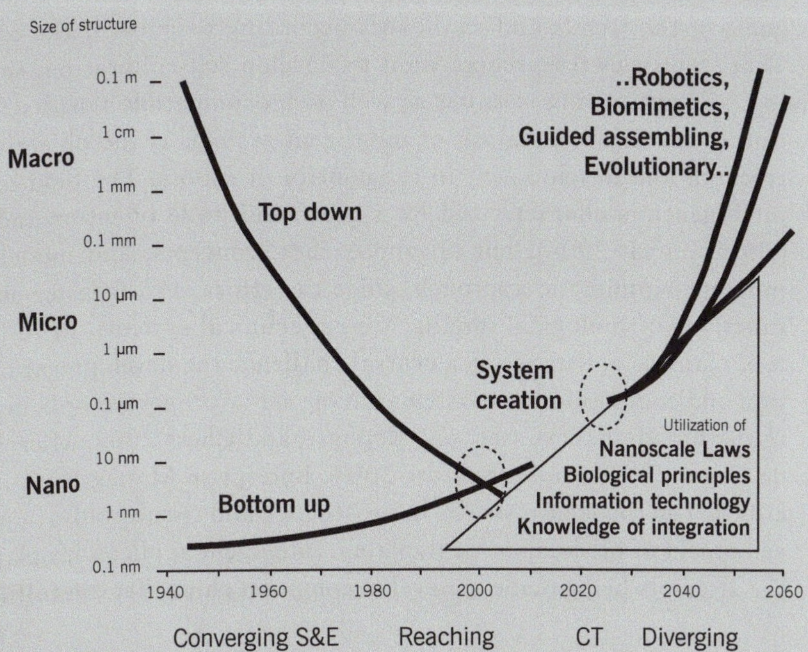

[Source: Roco 2005: 12]

Bottom-up vs. top-down approaches

sometime before 2020.[52] This intensified cooperation among the research and technology fields will make it possible to comprehend and master the structural principles of the nanoscale realm; this in turn will lead to the development of new individual disciplines, one of them being biomimetics, according to Roco. The Figure above (Fig. 3) is based on and expands on a similar diagram tracing the development of top-down and bottom-up technologies that has been frequently shown in other contexts because of the clear picture it presents; it should make this trend towards a new differentiation once more clear. Similar tendencies, however, have also been described in other papers looking towards the future (e.g. see Microsoft 2005).

Heavy influence of biomimetics on converging technologies?

The figure is based upon similar diagrams by the VDI. The VDI graphs served to demonstrate the nanotechnological trend leading to a fusion of different disciplines of science and also the bottom-up und top-down approaches. Roco considers the merger of the different disciplines to be a starting point for further directions of development; these will be greatly influenced by biomimetics/bionics concepts.

3.3 Biomimetics and technomimetics – opportunities and risks in converging technologies

Self-x-attributes as objective of future technologies

Smart, intelligent, self-organizing, self-healing,[53] error-tolerant, self-reproducing, biocompatible, etc. systems and applications – these are the goals of the trends and challenges according to most of the case studies. Thus, sensorics researchers want to develop self-calibrating, self-monitoring, self-optimizing sensors, as well as biocompatible sensors for application in medicine. Creation of intelligent systems is the objective of most research and development in the control of robots. The field of artificial intelligence is characterized by a great multitude of terms and concepts (Sin ák et al. 2004), but all approaches, concepts, and models have an inherent biomimetic approach, since the effort is to transfer entities and functions of biological intelligence to technical systems. In ICT research, too, many scientists see as a central challenge the development of information and communication systems having *self-x* properties or being adaptive in the broadest sense (see, e.g. Kephard and Chess 2003, Microsoft 2005, de Casto 2005, Müller-Schloer 2004, Enterprise Management Associates 2006). The focus of research in prosthetics and tissue cultivation is the development of biocompatible implants. Intelligent applications play a role in the attempts being made to develop computer chips that can fulfill memory

52 For a bibliometric review of the convergence hypothesis see Lieshout et al. (2005).

53 Or more generally: self-x properties

functions (Nsanze 2005) or interpret signals from the brain.[54] Within the material sciences, the effort to use knowledge about self-organisation and folding principles of peptides and proteins for the production of biocompatible materials plays an important role (Anderson et al. 2004). Research on smart materials (i.e., materials adapting their properties to ambient requirements) is mentioned in most of the technology foresight studies that were investigated.[55]

3

In nanobiotechnology this involves not only a move into the smallest dimensions, but simultaneously the establishment of an essential focus on the "production processes" of nature, i.e., in particular, the bottom-up processes, molecular self-organisation, etc. (Goodsell 2004). Associated with this are not only new technical possibilities, but essentially an entirely new production paradigm, which in some cases might be realized by means of bionanotechnology.

All these research approaches have in common a close relationship to biomimetics, because the natural processes and systems have those properties that are being desired increasingly and expected in technical systems. To illustrate the pending paradigm shift, Bogatyrev produced a typified representation of what was considered to be an ideal technical system vs. the properties of a biological system (see Table 6).

»Learning from nature« as promising approach in complex and dynamic environments

In some fields, learning from biological systems and possibly emulating them appears quite promising. Simultaneously it becomes clear that the challenges described cannot be met on the basis of a singular line of research and technology, but it will essentially require interdisciplinary, transdisciplinary, and multidisciplinary cooperation. For instance, a household robot that can transport a full glass safely through winding corridors and hand it over to a person but is not able to distinguish between milk and orange juice can only be used for limited purposes. The situation with prosthetics is similar: development of prostheses requires the close cooperation of physicians, material scientists, and IT specialists to create biocompatible limbs and organs that are equal to the "original from nature" with respect to function, controllability, and action speed. In IT and material development, in turn, nanotechnology increasingly plays a major role. This is exactly the trend Roco/Bainbridge called "converging technologies" (2002). With respect to the converging technologies, biomimetics plays a double role: it is an essential element of numerous problem-solving approaches (see Nordmann 2004, Lieshout et al. 2006, Silberglitt et al. 2006), and, as with the material sciences and nanotechnology, it is a model and a training ground for such convergence.

Biomimetics as model and learning platform for converging technologies

54 www.cyberkineticsinc.com

55 See particularly: Max-Planck-Gesellschaft (2001), Foresight (2000), NIMS (2005), Bensaude Vincent (2001).

Table 6 — Typification of an ideal technical system as compared to a biological system; based on Bogatyrev (2004)

Ideal technical system	Biological system
Simple structure	Complex structure
Everlasting or have necessary life length	Mortal
Easy to operate (deterministic)	Difficult to operate (stochastic)
Minimum use of resources	Maximum use of resources
Minimum waste production	Minimum waste production
Maximum capacity reserves	Availability in abundance
Easy to repair	Self-repairing
With different modes for different environments	Adaptivity
Automatic	Self-regulated
Reliable	Reliable

[Source: Bogatyrev (2004)]

»Human enhancement« as objective for converging technologies?

At times the debates dealing with the objectives of the converging technologies are quite controversial and heated. Roco/Bainbridge (2002) and RAND (Antón et al. 2001, Silberglitt et al. 2006) emphasize that the convergence of the five disciplines opens up, in particular, new opportunities to enhance human abilities ("human enhancement") and should therefore be taken advantage of. The authors of the EU study (Nordmann 2004) are much more cautious and more skeptical regarding a) the technical possibilities and b) the desirability of their realisation.

Biomimetics and biotechnology

The emulation of nature plays an important role within the research approaches that will be followed in the course of the converging technologies. However, it sometimes remains unclear to what extent truly biomimetic or just biotechnological approaches are being addressed. The example of spider silk can make this clearer. The fascination with the properties of spider silk and how it is "manufactured" in the organism of the spider is shared by all alike. For exponents of biomimetics, it stands as a guiding principle or model for adaptive production. With respect to an appropriate biomimetic manufacturing process for spider silk, opinions may differ considerably. The ambitious genetic engineering procedure being implemented, for example, by the company Nexia,[56] in which the gene for spider silk is implanted in goats with the objective of producing spider silk within the milk, is not considered a biomimetic approach by some biomimetics scientists.[57] What exactly a biomimetic production process should look like,

56 "We take a single gene from a golden orb-weaving spider and put it into a goat egg. The idea is to make the goat secrete spider silk into its milk." NYT June 2002.

57 See, for example, Benyus. As to the question of whether Thomas Scheibel's work group at TU München should have been admitted to the BioKoN network, is said to have been highly controversial.

however, is still unknown. However, the technological problems of the "hard" processes have also not all been solved (see e.g. Dickinson 1999).

Depth of intervention and potential risks

Questioning the quality, possible risks, side effects, and consequences of a manufacturing process is not unique to the example of spider silk. It goes hand in hand with the new possibilities of molecular design at the nanoscale and even more so with the convergence of nanotechnology and biotechnology and/or genetic engineering. The drastically increased depth of intervention into natural systems at different scales likewise leads to drastically increased synthetic power – and depth of intervention and synthetic power are obviously associated with greater potential risks and far-ranging undesirable (and unexpected) side effects and consequences, as we have already experienced during the development of synthetic chemistry and genetic engineering. This also may hold true for the very graphic example of production of artificial mother-of-pearl and other projects based on the model of biomineralisation: the intended production of mother-of-pearl, bone, cartilage, and tooth can take place in extremely different ways, from template-controlled in vitro crystallisation to biotechnological tissue engineering and in vivo genetic engineering. The technology assessment criterion for the consequences of an increased depth of intervention is the increased *impact potential* based on the achieved power over objects or systems – its effect, the manner in which it is implemented, and its dimension. The latter can be particularly seen in the anticipated length of the respective impact chains in space and time. Roughly said, it is about the difference between using an earthmover vs. a spade vs. a spatula, or the difference between splitting rocks vs. splitting atoms. The potential impact force of a technology can thus be a criterion in assessing the impact dimension to be anticipated.

Depth of intervention and implementation contexts

Depth of intervention can additionally be used as a criterion for a preliminary evaluation of the expectable form of impact. It makes a difference whether a drug only affects a part of the body or has a systemic effect (for example, a hormone). It makes a difference whether a technical intervention only affects the peripheral nervous system or the central one. It makes a difference whether environmentally foreign, persistent, and extremely mobile CFCs are set free in the atmosphere – or propane or butane, which photochemically decompose within seven days. The technology assessment criterion for depth of intervention is intended to help us determine the degree to which the target system is affected. In the process, it is important to observe where the intervention is being applied – to peripheral (or redundant) system elements or to the base structures and / or control structures?[58]

58 See Steinfeldt et al. 2007, 17ff., Petschow et al. (2005). For the background, see von Gleich (1989), for details von Gleich (1998/99). He offers a set of methods and criteria for, in essence, an ecologically-oriented technology and material assessment.

Figure 4 — Converging technologies and biomimetics

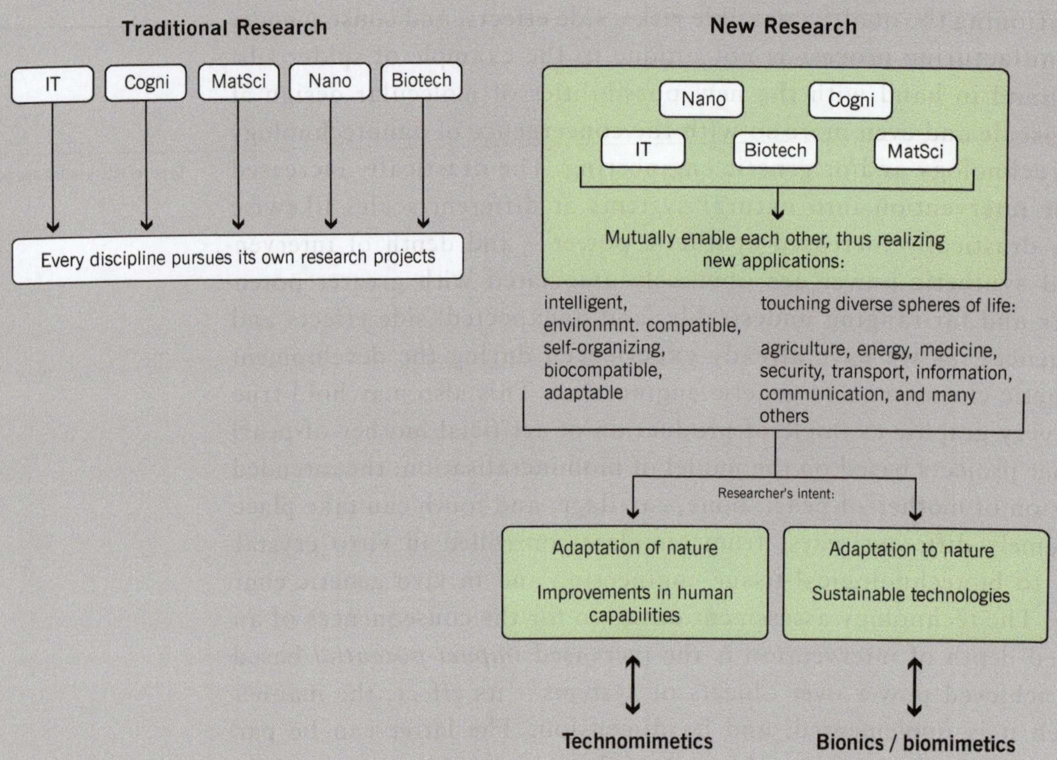

[Source: authors' own diagram]

Technomimetics

In the field of nanobiotechnology, too, scientists' objectives often extend far beyond the imitation of nature: nature serves not merely as a model for technological development; instead, a three-step process follows: (i) analysis of nature (molecular machines), (ii) analysis of the technical requirements of adaptation for industrial processes, (iii) adaptation of (the molecular machines of) nature to these technical production specifications. In this sense one could say that the first step involves biomimetics (technology based on the model of nature) and the third step technomimetics (emulated or modified biological systems based on the model of technology, see Petschow et al. 2005).[59]

The diagram depicts the different typified approaches that can be associated with the converging technologies. However, they cannot be derived directly from the various approaches mentioned, but on the other hand, can provide an orientation towards a direction of development and

59 In his article, Ball (2005) makes reference to the different directions of transfer: "more startling, perhaps, are possibilities such as programming cells to reproduce the algorithms of cellular automata – an ironic reversal of the metaphor – so that they interact with their neighbours in tightly prescribed ways, allowing them to develop spontaneous patterns, collective, and multicelled behaviour, and even forms of computing" (R7).

possibly also regarding depth of intervention. Finally, two different possible approaches can be distinguished, each likely to have very different consequences with respect to the associated application perspectives and risk dimensions: biomimetics, with biological solutions as its models, and technomimetics, which delves into the plasticity of biological systems in order to mechanize them (see, for example, approaches in synthetic and systemic biology). In both cases, the further technical solutions digress from the model of nature or a natural system of origin, the less the aspect of being "tried and proven by evolution" can be justified. It can be assumed that with respect to dealing with complexity, both directions of development will increasingly converge. Technical systems in their increasing complexity continue to move closer to the complexity of natural systems, and natural systems, with their necessarily reduced complexity required for conversion, move closer to that of technical systems. In both cases, we are in the end confronted with the practical problem of having to deal with complex and therefore not always fully understood systems – hence the need to reckon with unforeseeable impacts.

Is it justified to rely on solutions being evolutionary tried and tested?

3.4 Tentative conclusion regarding technological prospects

Smart, intelligent, self-organizing, self-repairing, error-tolerant, self-reproducing, and biocompatible etc. systems and applications are the objectives of the trends and challenges one finds when examining the case studies. Thus, sensorics researchers want to develop self-calibrating, self-monitoring, self-optimizing sensors, as well as biocompatible sensors for application in medicine. Creation of intelligent systems is the objective of most research and development in the control of robots. In ICT research as well, many scientists see the central challenge in developing information and communication systems having self-x properties or being adaptive in the broadest sense. The focus of research in prosthetics and tissue cultivation is the development of biocompatible implants. Intelligent applications play a role in the attempts being made to develop chips that can fulfill memory functions or interpret signals from the brain. Within the material sciences, the effort to use knowledge about self-organisation and folding principles of peptides and proteins for the production of biocompatible materials plays an important role. Research on smart materials (i.e. materials which adapt their properties to ambient requirements) is mentioned in most of the technology foresight studies that were studied.

Few explicit references to bionics or biomimetics, many direct references to »learning from nature«

Aim: implementation of self-x-attributes within new technologies

In nanobiotechnology this involves not only exploring the smallest dimensions, but simultaneously establishing an essential focus on the "pro-

New paradigm: bottom-up production processes

duction processes" of nature, i.e. in particular, the bottom-up processes, molecular self-organisation, etc. Associated with it are not only new technical possibilities, but essentially an entirely new production paradigm, which in some cases may be realised by means of bionanotechnology. This involves letting shapes and structures grow (e.g., by molecular self-organisation) instead of chiseling them out of a block or homogenising the material and casting or pressing it into a mold.

Biomimetics as model and training platform for converging technologies

Simultaneously it becomes clear that the challenges described cannot be coped with on the basis of a singular line of research and technology, but will require essentially interdisciplinary, transdisciplinary, and multidisciplinary cooperation. With respect to the converging technologies, biomimetics plays a double role: It is an essential element of numerous problem-solving approaches, and, as with material sciences and nanotechnologies, it is a model and a training ground for such convergence.

4 TOPICS, ACTORS, AND RESEARCH NETWORKS – RESULTS OF LITERATURE AND PATENT ANALYSES

»I find the ways that lead people to knowledge just as commendable as the nature of things itself.«

JOHANNES KEPLER, 1571–1630

4 TOPICS, ACTORS, AND RESEARCH NETWORKS – RESULTS OF LITERATURE AND PATENT ANALYSES

In this chapter the most significant results of the investigation of the international biomimetics research and development landscape are presented. The first section discusses the situation in Germany, whereas the latter sections take a European and global perspective. The network analysis results are further supplemented by the results of database analyses regarding publications and citation frequency. In accordance with the information that was gained from the technology foresight studies and the case studies (see chapter 3.2), we distinguished two categories in the analysis as well as in the following representation of the biomimetics research and development landscape: i) *biomimetics in the narrow sense* and ii) R&D approaches taking *learning from nature* as their basis.

See chapter 3.2

Biomimetics in the narrow sense

The first category includes those actors and institutions which explicitly acknowledge their involvement in biomimetics research and development (R&D), i.e., those conducting *biomimetics research in the narrow sense.* With respect to the content, these are R&D activities that are primarily rooted in the first (functional morphology) and second (signal and information processing) developmental strands in biomimetics (see chapter 2). *Biomimetics in the narrow sense* is also more likely to take place on the first (derived from the results of evolution) and second (derived from the evolutionary process itself) levels of learning from nature (see chapter 2).

See chapter 2

»Learning from nature«

The second category refers to those actors and institutions whose R&D approaches contain a clear reference to *learning from nature* as a component of possible technical solutions. In the majority of these cases, the R&D activities in this category can generally be assigned to the second and third developmental strands of biomimetics (molecular self-organisation and nanotechnology). *Learning from the (success) principles of nature* (third level; see chapter 2) is more widespread here than in *biomimetics in the narrow sense.* The second category, *learning from nature,* also covers a large number of those "biomimetic" approaches in which a merging of the three developmental strands is most evident and where a parallel learning from nature at all three levels (results of evolution, the evolutionary process, and success principles) takes place. The proximity to the "converging technologies" can also clearly be seen here.

See chapter 2

4.1 The national biomimetics R&D landscape

Analysis of biomimetics networks and institutions

A first approach to the German biomimetics landscape can be found in the analysis of the two major institutional research networks in Germany: "BioKoN" and "Kompetenznetz Biomimetik", which in their own words, "consist of the most important work groups in the areas of bionics" (BioKoN)[60] and constitute an "interface between [academic] research and industrial application" (Kompetenznetz Biomimetik).[61] Historically, before these two research networks, there was the "Gesellschaft für Technische Biologie und Bionik" (GTBB – "The Society for Technical Biology and Biomimetics"), which was founded in 1990.[62] An analysis of the membership structures, as well as participation in and contributions to conferences, was supplemented by interviews in which specific inquiries into the intensity of contacts and activities were made.[63] Thus, it was possible to identify and analyze the actors and institutions performing *biomimetics* research *in the narrow sense* (see above).

Interviews

4

Literature and patent analyses

Inquiries into the literature and journal databases provided a second approach in which we specifically searched for the terms "biomimetics" and "bionics" (using both the German terms and their English-language equivalents) as well as for terms that are frequently employed in describing approaches and concepts that either clearly are based on a learning from nature orientation or are at least not far removed (for example, self-organisation/self-healing). Thus, we were able, in particular, to identify actors and institutions not doing *biomimetics* research *in the narrow sense*, but in whose R&D activities, nonetheless, a clear reference to *learning from nature* can be found.

4.2 The German research networks "BioKoN" and "Kompetenznetz Biomimetik"

»Biomimetics Competence Network« (BioKoN)

Unlike the rather loose membership requirements of the GTBB, admittance to these two research networks is by formal application. By this, the application process of the two institutional biomimetics networks ensures a certain degree of quality assurance, too. The "Bionik-Kompetenz-Netz" (BioKoN – "Biomimetics Competence Network") was founded in 2001 as part of a project funded by the Federal Ministry of Education and

60 See: http://www.biokon.net/biokon/biokon.html (accessed: January 2007).

61 See: http://www.biologie.uni-freiburg.de/biomimetik/index.html (accessed: January 2007).

62 See: http://www.gtbb.org/ (accessed: January 2007).

63 Interview participants are listed on http://www.tecdesign.uni-bremen.de/bioniktrends. Also based on excerpts from informal conversations at workshops held in the course of this study. The list of participants is likewise on http://www.tecdesign.uni-bremen.de/bioniktrends.

Table 7 — Universities participating in the BioKoN and Kompetenznetz Biomimetik research networks

No.	University	Institution / department	Participants (direction)
1	RWTH Aachen	Bionik Zentrum Aachen (combines teaching and research at RWTH Aachen and is based at the Institut für Biologie II)	PD Dr. H. Luksch (managing director)
2	Technische Universität Berlin	Chair in Biomimetics und Evolutionary Optimisation Strategy	Prof. emerit. Dr. I. Rechenberg (former director)
3	Universität Bielefeld	Biological Cybernetics / Theoretical Biology Work Group (part of the Biology Faculty)	Prof. Dr. H. Cruse (director of the work group)
4	Universität Bochum	Sensory and Cognitive Ecology of Marine Mammals Research Group, Chair for General Zoology and Neurobiology (Chair for Biology)	PD Dr. G. Dehnhardt (director of research group)
5	Universität Bonn	Biomimetics Research Group at the Nees Institute for the Biodiversity of Plants; Department of Neuroethology and Sensory Ecology and the Department of Comparative Neurobiology (both in the Institute for Zoology)	Prof. Dr. W. Barthlott, Prof. Dr. G. von der Emde, Prof. Dr. H. Bleckmann (ea. director of research group / department)
6	Hochschule Bremen	Biomimetics subject area in the Department of Naval Architecture, Ocean Engineering, and Applied Sciences Bionik-Innovations-Centrum Bremen (BIC)	Prof. Dr. A. Kesel (director)
7	Technische Universität Darmstadt	Center of Biotechnical Engineering Darmstadt (BitZ) (structural framework of the biotechnology research focus at TU Darmstadt)	Prof. Dr. C. Tropea (spokesperson), Dr. H. Bargel (management)
8	Technische Universität Dresden	Chair in Botany (in the Institute for Botany)	Prof. Dr. C. Neinhuis (director)
9	Universität Freiburg	Plant Biomechanics Group of the Botanic Gardens of Universität Freiburg	Prof. Dr. T. Speck (director)
10	Universität Göttingen	Institute for Forest Zoology and Forest Conservation (in the Faculty of Forest Sciences and Forest Ecology)	Prof. Dr. S. Schütz (director)
11	Technische Universität Ilmenau	Biomechanics (in the Department of Mechanical Engineering)	Prof. Dr. H. Witte (director)
12	Fachhochschule Köln	Department of Information, Media, and Electrical Engineering	C. Hacker (adjunct professor)

Research areas	Teaching	Research network
in many diverse areas at 20 institutes (total of 6 faculties)	courses and lectures	BioKoN
in various fields, particularly flow optimisation, evolutionary optimisation strategy, robotics, and energy production	courses and lectures (lectures, seminars, internships)	BioKoN
in the areas of robotics (sensor technology, actuator engineering, control)	courses and lectures	BioKoN
on the orientation mechanisms of marine mammals	unknown	BioKoN
in the areas of surfaces (plant) and flow/ electro/infrared sensors (animals)	courses and lectures	BioKoN
in various fields, particularly shipbuilding, ocean engineering	international course of study in biomimetics (ISB)	BioKoN
in many diverse areas (total of 30 faculty members)	interdisciplinary area of specialisation in biotechnology	BioKoN
on surfaces/interfaces of plants	courses and lectures	BioKoN
in various areas related to plants (material development)	university research project	BioKoN/KN Biomimetik
in the area of insect sensors	unknown	BioKoN
in the areas of microsystems, medical technology, and robotics	area of specialisation biomechatronics	BioKoN
unknown	unknown	BioKoN

→ continuation next page

4

No.	University	Institution / department	Participants (direction)
13	Hochschule Magdeburg-Stendal	Industrial Design Institute (in Department of Design)	Prof. Dr. U. Wohlgemuth (director)
14	Hochschule Mannheim	Institute of Engineering Design/CAD and Institute of Biological Process Engineering (both in the Department of Chemical Process Engineering)	Prof. Dr. K. Peschges; Prof. Dr. P. M. Kunz (respective directors)
15	Technische Universität München	Fiberlab (in Department of Biotechnology in Division of Organic Chemistry and Biochemistry)	Dr. T. Scheibel (lab supervisor)
16	Universität Münster	Institute of Technics and Didactics (in Department of Physics)	Prof. Dr. B. Hill (director)
17	Hochschule Saarbrücken	Institute for Product Development and Optimisation (InPEO – made up of several faculty members and labs of HTW Saarbrücken) Department of Architecture	Prof. Dr. H.-J. Weber (contact person, InPEO); Prof. Dr. G. Pohl (Dept. of Architecture)
18	Universität des Saarlandes	BioKoN Saarbrücken site (office and coordination of BioKoN sites)	Prof. emerit. Dr. Nachtigall
19	Universität Tübingen	Functional Morphology and Biomimetics Project Group (at the Institute of Geosciences); Department of Evolutionary Biology of Invertebrates (at the Institute of Zoology)	Prof. Dr. V. Mosbrugger (former project group director); Prof. Dr. O. Betz (department director)
20	Universität Ulm	BioFuture Research Group (in the Department of Neurobiology)	Dr. F.-O. Lehmann (director)

[Source: own Internet enquiry]

Research (BMBF- "Bundesministerium für Bildung und Forschung") and was placed under the aegis of an officially registered association bearing the same name in 2004 (second phase of the project). The objective of BioKoN is "to demonstrate and fully explore the potentials of biomimetics in industry and science" as well as to create "a sustainable and permanent infrastructure for biomimetics" in Germany.[64] The BioKoN research network, with a total of four standing as well as seven thematic interest groups, covers a good deal of the biomimetics research spectrum (especially of *biomimetics in the narrow sense*).

»Competence Network Biomimetics« (KN-Biomimetik)

The "Competence Network Biomimetics: Plants as Concept Generators for the Development of Biomimetic Materials and Technologies" ("KN Biomimetik – Kompetenznetz Biomimetik") is an association of biomimetics researchers from the universities of Freiburg and Tübingen and the "Institut für Textil- und Verfahrenstechnik" (ITV – "Institute for Textile and

64 See the Web pages of BioKoN at: http://www.biokon.net/biokon/biokonev.html (accessed: January 2007).

Research areas	Teaching	Research network
biomimetic design	courses and lectures	BioKoN
unknown	unknown	BioKoN
biological fibers, particularly spider silk	unknown	BioKoN
biomimetics in school education (teacher training)	courses and lectures	BioKoN
various areas, particularly in biomimetic engineering and construction	unknown	BioKoN
unknown	unknown	BioKoN
project group: water transport in plants; department: unknown	unknown	BioKoN/KN Biomimetik
insect flight	unknown	BioKoN

Processing Engineering") Denkendorf.[65] It was founded in 2002 and has been supported since then by the "Ministerium für Wissenschaft, Forschung und Kunst des Landes Baden-Württemberg" (MWK – "Ministry of Science, Research, and the Arts Baden-Württemberg"). In contrast to BioKoN, it is both thematically narrower in its focus as well as organisationally much smaller.[66]

Results of network analyses

The above and the following tables present a summary of the information retrieved from the web sites of the two biomimetics research networks.[67] The results are organized in three groups: university institutions (Table 7), non-university R&D institutions (Table 8), and companies (Table 9). Initially, only those facilities and persons explicitly named as members or

65 See the Web pages of KN Biomimetik at: http://www.biologie.uni-freiburg.de/biomimetik/index.html (accessed: January 2007).

66 It is itself a member of BioKoN.

67 BioKoN: http://www.biokon.net/index.shtml /Kompetenznetz Biomimetik: http://www.biologie.uni-freiburg.de/biomimetik/.

cooperation partners in the networks were listed. A total of approximately twenty universities, nine non-university R&D institutions, and nine companies were identified. It thus becomes clear that both networks are (still) very much characterized by the universities, and that partners from industry scarcely appear directly, at least not as members or cooperation partners.

Universities in BioKoN and KN-Biomimetik

The quantity and intensity of the biomimetics activities at these twenty university institutions (Table 7) can be divided into roughly two equal categories.[68] First, there are those locations at which biomimetics also plays an extensive role in external communications, for example Technical University (TU) Darmstadt, home to its own "BiotechnikZentrum" (BitZ – "Center of Biotechnical Engineering Darmstadt") since 2003; TU Berlin, where the chair of "Biomimetics and Evolutionary Optimisation Strategy" has existed since 1972; or the "Hochschule Bremen" ("University of Applied Sciences Bremen"), with its "Bionik-Innovations-Centrum" (B-I-C – "Biomimetics Innovations Centre"), since 2004, and the world-wide first, and as far as we know to date only, full-term course of study in biomimetics (since 2003). In these centers we find various actors from various disciplines that have been involved with biomimetics research and teaching over a longer period of time. To this category of universities, with a relatively broad make-up in the field of biomimetics with respect to subject areas, personnel, and structure, we can further assign (with some restrictions or qualifications) the universities in Freiburg (KN Biomimetik) and Bonn ("Nees-Institut", lotus effect), TU Ilmenau (biomechanics), and RWTH Aachen ("Bionik-Zentrum Aachen").

Then, there is a second group of universities, where biomimetics is generally confined to one (or part of one) department or chair and only a limited number of people are involved. In this category we find the universities of Bielefeld (robotics) and Bochum (orientation mechanisms of marine mammals), TU Dresden (plant surfaces and interfaces), Universität Göttingen (sense organs of insects), Hochschule Magdeburg-Stendal (design), and Universität Münster (didactics). With additional restrictions or qualifications, we can also add the universities of applied sciences in Mannheim and Cologne as well as the universities in Munich (TU), Tübingen and Ulm.

»Society for Technical Biology and Bionics« (GTBB)

Yet to be mentioned is the Saar University in Saarbrücken, where Prof. Nachtigall has held the chair in zoology since 1969. He and his group became involved early on with biomimetics and, in 1990, founded the field of study "Technical Biology and Biomimetics" (which he supervised until his retirement in 2002) and also the already mentioned "Society for Technical Biology and Bionics" (GTBB). The initiative for the biennial "Biomimetics Confer-

68 Within the three categories that are subsequently described, there are also significant differences between the individual universities. Nonetheless, the subdivision into three categories makes sense, as it makes visible general tendencies.

Table 8 — Non-university public or non-profit R&D institutions participating in the BioKoN and Kompetenznetz Biomimetik networks

No.	Institution	Department(s)	Participants (direction)	Research network
1	Alfred-Wegener-Institut für Polar- und Meeresforschung Bremerhaven	Biological Oceanography; Plankton Biomechanics	Dr. Markus Geisen; Dr. Christian Hamm; Dr. Christof Baum	BioKoN
2	Forschungszentrum Karlsruhe GmbH,	Department of Biomechanics at the Institute for Material Research II	Prof. Dr. Claus Mattheck; Dr. Iwiza Tesari; Roland Kappel	BioKoN
3	Fraunhofer Institut für Fertigungstechnik und Angewandte Materialforschung, IFAM Bremen	Biopolymers section in the Department of Adhesive Bonding Technology and Surfaces (IFAM)	Dr. Klaus Rischka; Dr. Volkmar Stenzel; Dr. Ingo Grunwald	BioKoN
4	Fraunhofer Institut für Fabrikbetrieb und Fabrikautomatisierung (IFF) Magdeburg	Robot Systems	Dr. Norbert Elkmann; Dr. Oliver Lange	BioKoN
5	Fraunhofer Institut Umwelt-, Sicherheits-, Energietechnik UMSICHT Oberhausen	Advanced Materials Business Unit	Jürgen Bertling; Marcus Rechberger	BioKoN
6	Gesellschaft zur Förderung angewandter Informatik e. V. (GfaI), Berlin	Signal Processing	Dr. Ralf Schröder; Dr. Olaf Jaeckel	BioKoN
7	Institut für Textil- und Verfahrenstechnik (ITV) Denkendorf	Biomimetics Research Group	Prof. Dr. Heinrich Planck; Dr. Thomas Stegmaier; Dr. Markus Milwich	BioKoN/KN Biomimetik
8	Max-Planck-Institut für Metallforschung Stuttgart	Evolutionary Biomaterials Research Group	Dr. Stanislav Gorb; Dr. Dagmar Voigt	BioKoN
9	Institut für Technik und Design e.V., ITD Ingolstadt	unknown	Prof. Dr. Jörg Wellnitz; Michael Hartmann	BioKoN

[Source: own Internet enquiry]

ence" also came from Saarbrücken; the conferences were an essential element in the shaping of the identity of this research and development direction.[69] Prof. Nachtigall and Prof. Rechenberg (holding the chair of "Biomimetics and Evolutionary Optimisation Strategy" at TU Berlin, founded in 1972) –

69 Web site of the BioKoN location Saarbrücken: http://www.uni-saarland.de/fak8/bi13wn/index.html (accessed: January 2007).

Table 9 — Companies in the BioKoN network

No.	Company name	Field	Participants (direction)
1	EvoLogics GmbH, Berlin	numerous R&D activities in the area of biomimetics and bionics, underwater communication, flow control, fin ray	Dr. Rudolf Bannasch, Dr. Konstantyn Kebkal
2	inpro – Innovationsgesellschaft für fortgeschrittene Produktionssysteme in der Fahrzeugindustrie mbH, Berlin	R&D knowledge transfer in the area of automobile production	Dr. Michael Herdy
3	Dr. Brand Research & Consult, Berlin	unknown	Prof. Dr. Frank Brand
4	Architekturbüro Oligmüller, Bochum	design and development of buildings	Dieter Oligmüller
5	Deutsches Forschungszentrum für Künstliche Intelligenz (DFKI) GmbH, Kaiserslautern / Saarbrücken / Bremen	R&D activities in the area of robotics (Robotics Research Group)	Prof. Dr. Frank Kirchner; Dr. Dirk Spenneberg; Jan Albiez
6	Bionic StreamForm, Saarbrücken	prototype and model production for R&D purposes	Frank Wedekind
7	Dr. Mirtsch GmbH, Teltow	manufacturing of vault-structured sheet metals	Dr. Mirtsch
8	Suntec-Orga GmbH, Velten	manufacturing and marketing of a solar thermal system	Helmut Juran
9	Brodbeck Medienproduktion, Munich	production of media materials on the subject of biomimetics and bionics	Thomas Brodbeck

[Source: own Internet enquiry]

who was influenced by Prof. Hertel (aircraft construction) – dominated and characterized for many years – not least, with their publications and textbooks – (along with Dr. Mattheck, Prof. Heydemann, and later Prof. Barthlott) the appearance of the German biomimetics community (with respect to *biomimetics in the narrow sense*). Meanwhile, there is already a somewhat broader second generation of active biomimetics researchers, many already holding "positions of authority" (Bannasch, Cruse, Kesel, Speck, Mosbrugger, Neinhuis, Tropea, Witte); even so, it is still quite remarkable that the Saar University in Saarbrücken discontinued almost all their biomimetics activities with the retirement of Prof. Nachtigall. The appointment of someone to fill Rechenberg's chair at TU Berlin has also been delayed.[70]

In Table 8 we included those non-university R&D institutions registered with BioKoN or KN Biomimetik that are (were) publicly financed or oper-

70 The departments at Technical University of Berlin apparently could not agree on which would host the successor. This would be a rather not untypical problem given interdisciplinary research areas.

Extra-faculty research and development institutions in BioKoN and KN-Biomimetik

ate as non-profit institutions (e.g., registered societies and associations).[71] A third group of institutions that does research and development and is organized as limited liability companies (GmbH) or in a comparable legal form was categorized by name of (parent) company and listed in Table 9. The allocation to these tables makes clear, once more, that the tendency for biomimetics in Germany is to be driven by the research groups – technology push – and less so, demand. The two biomimetics networks are dominated by the public or non-profit institutions; business is represented only weakly.

In the non-university R&D institutions the intensity of the biomimetics activities ranges from individual departments or research groups (for example the „Bionics Research Group“ at ITV Denkendorf, the „Evolutionary „Biomaterials Research Group“ at the Max-Planck-Institute for Metals Research, and the „Department of Biomechanics“ at the Institute for Materials Research II at the Research Centre Karlsruhe) to isolated activities (for example, the Institute for Technology and Design in Ingolstadt or the Fraunhofer Institute for Factory Operation and Automation IFF). However, it should have been mentioned already that the biomimetics research activities of the Max Planck Institute (MPI) of Colloids and Interfaces in Golm are at least as intensive as those at the MPI in Stuttgart. There are also clear indications that MPI Golm is moving closer to the BioKoN research network.

Enterprises in BioKoN and KN-Biomimetik

Only those companies listed as *members* of BioKoN are included in Table 9.[72] Two-thirds of the companies mentioned are small entrepreneurial businesses or microbusinesses; the remainders are small businesses, and in some cases only a few employees are involved in biomimetics (as, e.g., in case of inpro). This does not, however, say anything about possible cooperative involvements with companies not beeing part of the research network.

4.3 GTBB, conferences, and expert interviews

Wide context of biomimetics actors

In the following tables (Tables 10 through 12) we summarize the results of the analysis of the memberships in the German Society for Technical Biology and Bionics (GTBB), the conferences, and the expert interviews.

In Table 10, with a total of approximately 60 entries, there are three times as many universities with biomimetics connections as compared to the analysis of the two research networks (see above). However, the majo-

71 However, the Fraunhofer institutes, with their modest base funding and large contributions from industry, like the ITV Denkendorf are not classified in the same way.

72 KN Biomimetik allows no business memberships.

Table 10 — Universities in Germany with biomimetics activities

No.	University	Institution / department	Participants (direction)
1	FH Aachen	Department of Air and Space Technology	Prof. Dr.-Ing. Mertens
2	RWTH Aachen	Bionics Center Aachen	PD Dr. H. Luksch (managing director)
3	FH Aalen	European Center of Excellence Metal Casting Fachhochschule Aalen	Prof. Dr. Klein; Prof. Dr. Waidmann
4	Universität Augsburg	Chair of System-Oriented Information Technology and Communications Systems	Prof. Dr. Ungerer
5	FU Berlin	Department of Mathematics and Informatics; Institute for Neurobiology	Prof. Dr. Raul Rojas; Prof. Dr. Menzel
6	HU Berlin, Charité, Universitätsmedizin Berlin	Biofluidmechanics Lab; Clinic of General, Visceral, and Transplantation Surgery	Dr.-Ing. Ulrich Kertzscher; Prof. Dr.-Ing. Klaus Affeld; Dr. med. Sauer
7	TU Berlin	Chair of Biomimetics and Evolutionary Optimisation Strategy	Prof. emerit. Dr. I. Rechenberg
8	Universität Bielefeld	Chair of Biological Cybernetics / Theoretical Biology	Prof. Dr. H. Cruse
9	Universität Bochum	Biomechanical Constructions Research Group; Photobiotechnology Work Group; Institut für Neuroinformatik (INI); Sensory and Cognitive Ecology of Marine Mammals Work Group	Dr.-Ing. Witzel; Prof. Dr. T. Happe; Dr. C. Igel; Prof. Dr. G. Dehnhardt
10	Universität Bonn	Nees-Institute for Plant Biodiversity; Neuroethology and Sensory Ecology; Comparative Neurobiology	Prof. Dr. W. Barthlott; Prof. Dr. von der Emde; Prof. Dr. Bleckmann; PD Dr. Schmitz
11	Hochschule Bremen	Biomimetics	Prof. Dr. A. Kesel
12	Hochschule für Künste Bremen	Institute of Integrated Design	Prof. Dr.-Ing. Kraus
13	Jacobs University Bremen (formerly International University Bremen)	Biochemical Engineering; Biophysics	Prof. Dr. Ulrich Schwaneberg; Prof. Dr. Mathias Winterhalter
14	Universität Bremen Dep. Production Engineering	Department of Ceramic Materials and Components, Technological design and development; Bremen Institute of Industrial Technology and Applied Work Science (BIBA); Institute for Microsensors, Actuators, and Systems (IMSAS)	Prof. Dr.-Ing. Georg Grathwohl, Prof. Dr. Arnim von Gleich; Prof. Dr.- Ing. Bernd Scholz-Reiter; Prof. Dr.-Ing. Walter Lang

No.	University	Institution / department	Participants (direction)
15	TU Clausthal	Institute of Non-Metallic Materials	Prof. Dr. Jürgen G. Heinrich; Dr. Lothar Wondraczek
16	TU Darmstadt	Center of Biotechnical Engineering Darmstadt (BitZ)	Prof. Dr. C. Tropea (spokesperson), Dr. H. Bargel (management)
17	Universität Dortmund	Department of Logistics	Dipl.-Ing. Katja Nowka
18	TU Dresden	Institute of Waste Management and Contaminated Site Treatment; Institute of Conveying Technology, Construction Machines and Logistics; Institute of Steel and Timber Construction; Institute of Conveying Technology, Construction Machines and Logistics; Chair in Botany	Prof. Dr. Peter Werner; Dr.-Ing. Frank Schulze; Prof. Dr.-Ing. Peer Haller; Dipl.-Ing. Beate Buchelt; Prof. Dr. C. Neinhuis
19	Universität Duisburg-Essen	Institute of Product Engineering	Prof. Dr.-Ing. Peter Köhler
20	Universität Erlangen Nürnberg	Institute of Material Sciences (Glass and Ceramics)	Dr. Frank A. Müller
21	TU Freiberg	Institute of Mechanics and Fluid Mechanics	Prof. Dr.-Ing. Brücker
22	Universität Freiburg	Botanic Garden, Kompetenznetz Biomimetik; Chair in Construction of Microsystems	Prof. Dr. Thomas Speck; Dr.-Ing. Elmar Just; Dr.-Ing. Norbert Kockmann
23	FH Furtwangen	Department of Mechanical and Process Engineering	Prof. Dr.-Ing. Tawakoli; Prof. Dr.-Ing. Kubisch
24	Universität Giessen	Institute of Applied Physics	PD Dr. von Kreutzbruck
25	Universität Göttingen	Institute for Forest Zoology and Forest Conservation	Prof. Dr. S. Schütz
26	Universität Halle-Wittenberg	Institute for polymer materials (IPW); Institute for plant breeding and protection; Institute of Bioengineering	Dipl.-Phys. Sven Henning; Prof. Dr. Holger B. Deising; Prof. Dr. Markus Pietzsch
27	Universitätsklinikum Hamburg-Eppendorf	Policlinic for orthodontics	PD Dr. Klocke
28	Medizinische Hochschule Hannover (MHH)	Department of Rheumatology	Jens G. Kuipers
29	Universität Hohenheim	Institute for Environmental and Animal Hygiene	Prof. Dr. Böhm

→ continuation next page

No.	University	Institution / department	Participants (direction)
30	TU Ilmenau	Biomechatronics	Prof. Dr. H. Wurmus; PD Dr.-Ing. L. Zentner; Dr.-Ing. E. Kolev; Dr. C. Schilling; Prof. Dr. H. Witte
31	Universität Jena	Institute of Biochemistry II; Locomotion Laboratory	Dr. Jörg -Hermann Ozegowski; Dr. Seyfarth
32	TU Kaiserslautern	Institute of Integrated Sensor Systems; Robotics Research Lab (Dept. of Computer Sciences),	Prof. Dr.-Ing. König; Prof. Dr. Berns
33	Universität Karlsruhe	Institute for Applied Computer Science (AIA)	Dipl.-Ing. Thorsten Röllich
34	Universität Kassel	Field of expertise: Science of Structural Design ; Department of Turbomachinery	Univ.-Prof. Dr.-Ing. Martin Lawerenz
35	Universität Koblenz-Landau	Center for Human Resource Management	Dr. Martina Sailer
36	FH Köln	Faculty of Information, Media, and Electrical Engineering	C. Hacker
37	Universität Leipzig	Institute of Building Design and Management; Saxon Institute of Applied Biotechnology; Institute of Anatomy,	Prof. Dipl.-Ing. Architekt Bukhard Pahl; Dr. J. Ondruschka; Dr. Thomas Wiener
38	Universität Lüneburg	Computer Engineering	Prof. Dr.-Ing. Welge
39	FH Magdeburg-Stendal	Dept. of Engineering and Industrial Design	Prof. Dr. Wohlgemut
40	Hochschule Mannheim	Institute of Biological Process Engineering (IBV); Institute of Engineering Design/CAD	Prof. Dr. Kunz; Prof. Dr. K. Peschges
41	FH München	Dept. of Electrical Engineering and Information Technology	Prof. Dr. Wassermann
42	TU München	Chair of Food Packaging Technology; Chair of Biotechnology	Dr.-Ing. Kajetan Müller; Dr. Scheibel
43	Universität der Bundeswehr München	Institute Engineering Mechanics and Structural Mechanics	Prof. Dr.-Ing. Norbert Gebbeken
44	FH Münster	Fluid Mechanics Laboratory	Prof. Dr.-Ing. Jantzen
45	Universität Münster	Institute of Technics and Didactics	Prof. Dr. B. Hill
46	Universität Potsdam	Analytical Biochemistry	Dr. Axel Warsinke
47	FH Reutlingen	Institute of Applied Research	Prof. Dr. Gerhard Schulz
48	Universität Rostock	Institute of Applied Microelectronics	Prof. Dr.-Ing. habil. Ralf Salomon

No.	University	Institution / department	Participants (direction)
49	Hochschule Saarbrücken	Institute for Product Development and Optimisation (InPEO); Department of Architecture	Prof. Dr. H.-J. Weber; Prof. Dr. G. Pohl
50	Universität des Saarlandes, Saarbrücken	BioKoN Saarbrücken	Prof. emerit. Dr. W. Nachtigall
51	FH Schmalkalden	Business Sciences	Prof. Dr. Friedrich Kugler
52	Universität Stuttgart	Institute for Human Factors and Technology Management (IAT); Institute for Machine Tools	Dr.-Ing. A. Bröcker; Prof. Dr.-Ing. habil. J. Tröger; Dipl.-Ing. S. Ivanova
53	FH Südwestfalen	Electrical Energy Technology	Dr.-Ing. Berthold Bitzer
54	Universität Tübingen	Center for Bioinformatics (ZBIT); Institute for Physical and Theoretical Chemistry; Institute of Anatomy; Institute of Geoscience (IFG); Institute of Organic Chemistry; Wilhelm Schickard Institute	Dipl.-Ing. F. Strechert; PD Dr. U. Weimar; Dipl.-Ing. T. Schmidt; Dr. A. Roth-Nebelsick; Dr. B. Görlach; Dipl.-Phys. A. Lilienthal
55	Universität Ulm	Biomimetic and Bioinorganic Chemistry; Institute for Neurobiology;	Prof. Dr. Dirk Volkmer; Dr. Fritz-Olaf Lehmann
56	Universität Weimar	Hydrolab Schleusingen	Dr. Kranawettreiser
57	Internationales Hochschulinstitut Zittau	Environmental Process Engineering	Dr.-Ing. Kayser
58	Westsächsische Hochschule Zwickau	Textile and Leather Engineering	Prof. Dr. Hessberg

[Source: own Internet enquiry]

rity of the universities listed are not actively involved on an ongoing basis in biomimetics R&D. At many universities, only a single project related to biomimetics could be identified.[73]

There are two possible reasons for some researchers or research groups at universities in Table 10 not being listed in Table 7. First – and this may be true in the majority of the cases – biomimetics activities, at least to date, are not being carried out on an ongoing basis. BioKoN and KN Biomimetik offer a fairly accurate picture – *qualitatively* seen, at least – of those universities that are explicitly carrying out biomimetics research and teaching beyond the isolated single project.

73 In the case of a few universities, it was unclear despite inquiry, which activities were being (or had been) carried out there in the area of biomimetics; even so, they were included in the table, as they had been named by one of the sources mentioned. A detailed and in-depth analysis, i.e., comprehensive appraisal of the individual research and teaching activities and conversations with university faculty members at all 60 universities was not possible within the scope of this project.

Table 11 — Non-university R&D institutions in Germany with biomimetics activities

No.	Institution	Department(s)	Participants (direction)
1	Alfred-Wegener-Institut für Polar- und Meeresforschung, AWI, Bremerhaven	Biological Oceanography; Plankton Biomechanics;	Dr. Markus Geisen; Dr. Christian Hamm;
2	Deutsches Zentrum für Luft- und Raumfahrt e.V., DLR,	Institute of Composite Structures and Adaptive Systems, Braunschweig; Institute of Propulsion Technology, Köln / Berlin / Göttingen	Dr. Jörg Melcher; Dr. Wolfram Hage
3	Faserinstitut Bremen e. V., FIBRE	Field of expertise: Biobased Fiber re-enforced Materials / Sustainability	Dr.-Ing. Jörg Müssig
4	Forschungsgemeinschaft Bionik-Kompetenznetz e.V., BioKoN, Berlin	-	Dr. Rudolf Bannasch
5	Forschungsinstitut für Tief- und Rohrleitungsbau Weimar e.V.	Department of Biotechnology / Bionics	Dr.-Ing. W. Berger
6	Forschungszentrum Karlsruhe GmbH, FZK	Institute for Applied Computer Science (AIA); Institute of Nanotechnology (INT); Institute for Materials Research II	Dr. Ing. Stefan Schulz; PD Dr. Teodor S. Balaban; Prof. Dr. Claus Mattheck
7	Fraunhofer-Institut für Molekularbiologie und Angewandte Ökologie, IME, Aachen	Functional and Applied Genomics	Dr. Dirk Prüfer
8	Fraunhofer Institut Biomedizinische Technik, IBMT, St. Ingbert	Miniaturized Systems Work Group; Microsystems / Laser Medicine Work Group; Magnetic Resonance Work Group;	Dipl.-Ing. H. Andresen; Prof. Dr. H. Zimmermann; Dr. P. Geggier; Dr. O. Scholz; Dr. I. Riemann; Dr. B. Manz
9	Fraunhofer Institut Chemische Technologie, ICT, Pfinztal	(department dissolved)	Dipl.-Ing. Günter Helferich
10	Fraunhofer Institut Fertigungstechnik Materialforschung, IFAM, Bremen	Adhesives and Polymer Chemistry; Cellular Metallic Materials	Dr. Klaus Rischka; Dr. Volkmar Stenzel; Dr. Andreas Hartwig; Dr. Günter Stephani
11	Fraunhofer Institut für Betriebsfestigkeit und Systemzuverlässigkeit, LBF, Darmstadt	Demonstration Center SIMTOP	Dipl.-Ing. Klaus Störzel

No.	Institution	Department(s)	Participants (direction)
12	Fraunhofer Institut für Fabrikbetrieb und Fabrikautomatisierung, IFF, Magdeburg	Robotic Systems; Virtual Engineering	Dr. N. Elkmann; Dr. O. Lange; Dr. sc. techn. U. Schmucker
13	Fraunhofer Institut für Graphische Datenverarbeitung, IGD, Darmstadt	Visualisation and Virtual Reality	Dr. Ulrich Bockholdt
14	Fraunhofer Institut für Grenzflächen- und Bioverfahrenstechnik, IGB, Stuttgart	Biomimetic Interfaces; Pharmaproteins and Target Screening	Dr. Günter Tovar; Dr. Steffen Rupp
15	Fraunhofer Institut für Produktionsanlagen und Konstruktionstechnik, IPK, Berlin	Automation and Robotics	Dipl.-Ing. Jelena Radojicic;
16	Fraunhofer Institut Produktionstechnik und Automatisierung, IPA, Stuttgart	Robotic Systems; Technical Information Processing; Innovation Management	C. Meyer; Dipl.-Ing. H. Staab; Dipl.-Phys. J. Pannekamp; Dipl.-Phys. K. Drachsler
17	Fraunhofer Institut Umwelt-, Sicherheits-, Energietechnik, UMSICHT, Oberhausen	Advanced Materials	Dipl.-Ing. Jürgen Bertling; Dipl.-Ing. Marcus Rechberger
18	Fraunhofer-Institut für Physikalische Messtechnik, IPM, Freiburg i. Br.	Process Monitoring (Sensor Technology)	Dr. Jürgen Wöllenstein
19	Fraunhofer-Institut für Solare Energiesysteme, ISE, Freiburg i. Br.	Solar Thermal Systems	Dipl.-Ing. M. Hermann
20	Fraunhofer-Institut für Techno- und Wirtschaftsmathematik, ITWM, Kaiserslautern	Flows and Complex Structures	Priv.-Doz. Dr. Arnulf Latz
21	Gesellschaft zur Förderung angewandter Informatik e. V. (Gfal), Berlin	Signal Processing	Dr. Ralf Schröder; Dr. Olaf Jaeckel
22	Hahn-Meitner-Institut, HMI, Berlin	Solar Energy	Prof. Dr. Helmut Tributsch
23	Leibniz-Institut für Agrartechnik Bornim e.V., ATB	Agricultural Technology	Dr.-Ing. Hans-R. Langner
24	Institut für Mikro- und Informationstechnik der Hahn-Schickard-Gesellschaft e. V., HSG-IMIT, Villingen-Schwenningen	Microdosage Systems	Dipl.-Ing. Thorsten Göttsche
25	Institut für Physikalische Hochtechnologie e. V., IPHT, Jena	Biotechnical Microsystems Work Group	Dr. rer.nat. Wolfgang Fritzsche

→ continuation next page

4

No.	Institution	Department(s)	Participants (direction)
26	Institut für Textil- und Verfahrenstechnik (ITV) Denkendorf	Bionics Research Group	Dr. Thomas Stegmaier; Dipl.-Ing. Michael Linke; Dr. Markus Milwich
27	Max-Planck Institut für Kolloid- und Grenzflächenforschung, MPI-KG, Potsdam	Department of Biomaterials	Dr. Ingo Burgert
28	Max-Planck Institut für terrestrische Mikrobiologie, MPI-TM, Marburg	Department of Organismic Interactions	Dr. Michael Feldbrügge
29	Max-Planck-Institut für Metallforschung, Stuttgart	Evolutionary Biomaterials Group	Dr. Ulrike Wegst; Dr. Stanislav Gorb; Dr. Pablo Perez-Goodwynn
30	Nieklitzer Ökologie- und Ökotechnologie-Stiftung	Founder and Chairperson	Prof. Dr. Bernd Heydemann
31	Umweltforschungszentrum Leipzig-Halle GmbH, UFZ	Groundwater Purification	Dr. Arno Kaschl
32	Bundesanstalt für Materialforschung und –prüfung, BAM, Berlin	Materials and Environment	Dr. Kemmlein; Dr. Medianpour
33	Deutsches Krebsforschungszentrum, DKFZ, Heidelberg	Functional Genome Analysis	Dr. Frohme
34	Papiertechnische Stiftung München	Paper Ceramics	Dipl-Ing. Kirmeier
35	Sächsisches Textilforschungsinstitut e.V., STFI, Chemnitz	Warp-Knitting / Weaving	Mählmann
36	Stiftung caesar, Bonn	Biotechnology	Dr. Cerboncini

[Source: own Internet enquiry]

Identification of biomimetics actors

Nonetheless, there are also research groups at the universities that conduct ongoing research utilizing biomimetic approaches but are not members in either of the two networks. Two examples should be mentioned here in brief. Holder of the chair in biomimetic and bioorganic chemistry, Prof. Dirk Volkmer, at the University of Ulm describes its core assignment and the content of its work as being "biomimetic catalysis and bio-inspired material synthesis."[74] However, only the group led by Prof. Fritz-Olaf Lehmann at the Institute for Neurobiology at the University of Ulm, which uses the flight of insects as its point of origin for technical implementations, is neither a member of BioKoN and thus presented in Table 7. Volkmer's group (also at the University of Ulm), which is working at the molecular biological and chemical level on the transfer of biological findings into technical applications, is not a member of BioKoN nor of KN Biomimetik and thus missing in Table 7. And University of Augsburg only appears in Table 10.

74 Web site of the Chair of Biomimetic and Bioinorganic Chemistry: http://www.uni-ulm.de/ac2/Volkmer_Group/ (accessed: January 2007).

The group under the chair in "Systems and Networking" at University of Augsburg is doing research and development based on the model of nature, dealing with "organic computing" with a focus on biological principles and features such as "self-healing" and "self-optimisation." Such a biomimetics subject matter is also part of their teaching curriculum as the "Bio-Inspired & Grid Computing" demonstrate.[75] Thus, the University of Augsburg could easily become a member of one of the biomimetics research networks, yet to date it appears that no effort in this direction has been made.

Various self-definitions of biomimetics actors

The various actors in the area of biomimetics also differ in how they view themselves. With respect to this variation in self-image, three groups can be distinguished. Some persons/institutions – especially those that are organized in the research networks – view themselves as "biomimeticists" *("Bioniker")*, i.e., biomimetics researchers, and categorize their work under this term *(biomimetics in the narrow sense)*. With the actors of the second group the term *biomimetics* never appears and *biomimetically* or *biologically inspired* are used very rarely. These actors work in areas such as "organic computing" or "artificial photosynthesis," in which the concept of *learning from nature/biology* is already inherent. Finally, the members of the third group view themselves first and foremost as biologists, engineers, material science researchers, and chemists, etc., and refer to themselves as such. At the same time, however, this third group is in fact doing biomimetic and/or biologically inspired research and development, sporadically or even as their main focus. The fact that a form of learning from nature/biology played a role in this work is made explicit every now and then, but it remains in the background (borderline cases between the two categories, *biomimetics in the narrow sense* and *learning from nature*).

In comparison with Table 8, four times as many non-university R&D institutions that are or were involved in biomimetics activities are listed in Table 11. But here, too, at these additional non-university R&D facilities *biomimetic* research and development *in the narrow sense* and/or *learning from nature* is often only being carried out sporadically or to a limited extent.

Among the companies, the discrepancy between the actors represented in the research networks and those that were identified by means of conferences, expert interviews, and the GTBB is the greatest (a ration of 9 to 75). However, some companies were identified by means of an Internet search, the results of which at least presume that these companies relate positively to biomimetics approaches and possibly advertise this relation. How strong the actual commitment to or engagement in biomimetics is, remains unclear.

75 See the Web site of the Chair of System-Oriented Information Technology and Communications Systems: http://www.informatik.uni-augsburg.de/lehrstuehle/swt/vs/lehre/WS_06_07/seminar/ (accessed: January 2007).

Table 12 — Companies in Germany involved in biomimetics

No.	Company	Sector	Biomimetics Classification	Biomimetics Principle	Stage of Development (year)
1	3M Deutschland	chemicals	structure, form, and function*	shark skin	prototype / demonstration facility (1996)
2	Adam Opel AG	automotive	structure, form, and function*, locomotion, transport**	composites, fluid dynamics	prototype / demonstration facility
3	Airbus	aeronautics	structure, form, and function*	composites, shark skin	prototype / demonstration facility
4	Alan Sportartikel GmbH	sporting goods	structure, form, and function*	form	market introduction
5	Armstrong DLW AG	construction	structure, form, and function*	lotus effect	-
6	Audi AG	automotive	structure, form, and function*	biomimetic design	prototype / demonstration facility
7	BASF	chemicals	structure, form, and function*	lotus effect	-
8	Bionic Solutions	business consulting	extra***	business consulting	-
9	Bionic StreamForm	business consulting	locomotion, transport	fluid dynamics	-
10	Bionicon-Inwall GmbH	bicycle	structure, form, and function*	form	-
11	Bionik Design	media	extra***	media	-
12	BIONIK Media GmbH	media	extra***	media	-
13	Bionik-Systeme Bremen	power line construction	structure and form	river meander	market introduction
14	BMT Bionic Maritime Technologies GmbH	electronics/ electro-technology	sensor technology / information processing	sensor technology	market introduction
15	BMW	automotive	structure, form, and function*, locomotion, transport**	fluid dynamics, composite structures, lotus effect	market penetration

No.	Company	Sector	Biomimetics Classification	Biomimetics Principle	Stage of Development (year)
16	Chendo GmbH	medical technology	prosthetics, substances, and materials	composites, shark skin	-
17	ComNac - Communication Networks Aachen GmbH	software	optimisation	efficient information searches	-
18	Continental	automotive	structure, form, and function*	winter tire polar bear paws	market penetration
19	Daimler Chrysler	automotive	structure, form, and function*, locomotion, transport**	fluid dynamics, composites	prototype / demonstration facility
20	Degussa AG/ Degussa Creavis Technologies & Innovations	chemicals	structure, form, and function	lotus effect	market penetration
21	Dr. Mirtsch GmbH	metal production and processing	substances and materials	self-organisation vault structures, stiffening	-
22	Dr. Sander & Associates Software GmbH	software	optimisation, system design (industrial ecology)	swarm intelligence	-
23	Duravit AG	sanitary wares	structure, form, and function*	lotus effect	market introduction
24	EADS Deutschland GmbH	aeronautics	structure, form, and function*	lightweight construction	market introduction
25	Erlus	ceramics	structure, form, and function*	lotus effect	-
26	EvoLogics	sensor technology, underwater communication, robotics	sensors, structure, form, function	fluid dynamics, robotics, sensor technology	trials / market introduction?

→ continuation next page

No.	Company	Sector	Biomimetics Classification	Biomimetics Principle	Stage of Development (year)
27	EWABO Chemikalien GmbH & Co.KG	chemicals	structure, form, and function*	lotus effect	-
28	Ferro Corporation	chemicals	structure, form, and function*	lotus effect	prototype / demonstration facility
29	Festo AG & Co. KG	electronics / electro-technology	robotics	artificial muscle	-
30	Franz Kaldewei GmbH & Co. KG	sanitary wares	structure, form, and function*	lotus effect	market introduction
31	Friedrich Graepel AG	metal production and processing	structure, form, and function*	form	market penetration
32	Fumedica GmbH	medical technology	structure, form, and function*	lotus effect, shark skin	market introduction / market penetration
33	GFal	electronics / electro-technology	sensor technology and information processing	sensor technology	market penetration
34	Goldschmidt AG (Degussa Gruppe)	chemicals	structure, form, and function*	lotus effect	market introduction
35	Heidenreich & Harbeck AG	machine-building, foundry construction	optimisation	bio-algorithm	-
36	Ideal Standard	sanitary wares	structure, form, and function*	lotus effect	-
37	Innovent GmbH	construction	structure, form, and function*	lotus effect	-
38	Institut für Bionik	business consulting	extra***	business consulting	-
39	Interton Hörgeräte GmbH	medical technology	sensor technology and information processing	sensor technology	market penetration
40	Keramag AG	sanitary wares	structure, form, and function*	lotus effect	-
41	Kleinmann GmbH	cleaning products	structure, form, and function*	lotus effect	market introduction

No.	Company	Sector	Biomimetics Classification	Biomimetics Principle	Stage of Development (year)
42	Koralle-Sanitärprodukte GmbH	sanitary wares	structure, form, and function*	lotus effect	market introduction
43	KUKA Roboter GmbH	machine-building	robotics, sensor technology and information processing	sensor technology, robotics	market penetration
44	Living Systems GmbH absorbed into Whitestein Technologies	software	system design (industrial ecology)	swarm intelligence	market introduction
45	LTG Aktien-gesellschaft	construction	energy conversion and use	ventilation	market penetration
46	Miele & Cie. KG	electro-devices	substances and materials	vaulted structures	market penetration
47	M-PORE GMBH	metal production and processing	substances and materials	lightweight construction	-
48	Nano Maxi	nanotechnology	structure, form, and function*	lotus effect	market introduction
49	Nano X GmbH	nanotechnology	structure, form, and function*	lotus effect	-
50	NanoConcept	nanotechnology	structure, form, and function*	lotus effect	market introduction
51	NanoScape AG	nanotechnology	structure, form, and function*	lotus effect	market introduction
52	NuTech Solutions	business consulting	extra***	business consulting	-
53	Okalux	construction	structure, form, and function*	transparent thermal insulation (glass) utilizing the polar bear principle	market introduction / market penetration
54	Össur Europe	medical technology	prosthetics	prosthetics	market penetration
55	Otto Bock Healthcare GmbH	medical technology	prosthetics	prosthetics	market penetration

→ continuation next page

4

No.	Company	Sector	Biomimetics Classification	Biomimetics Principle	Stage of Development (year)
56	Percenta Europe Ltd.	nanotechnology	structure, form, and function*	lotus effect	-
57	Phonak GmbH	medical technology	sensor technology/ information processing, structure, form, and function*	sensor technology, lotus effect	market penetration
58	Pohl Architekten	architecture	structure, form, and function*, energy conversion and usage	lightweight construction, ventilation	-
59	Ravensburger	media	extra***	media	-
60	Rheinzink	construction	substances and materials	vaulted structures	market penetration
61	Rittal GmbH & Co. KG	construction	structure, form, and function*	lotus effect	market introduction
62	Schöller Textil AG	textiles	structure, form, and function*	lotus effect, function	market introduction
63	Siemens AG	electronics / electro-technology	sensor technology/ information processing, system design (industrial ecology)	sensor technology, swarm intelligence	market introduction
64	Siteco Beleuchtungs-technik GmbH	lighting	substances and materials	vaulted structures	market penetration
65	SolarEnergie Stefanakis	solar energy	energy conversion and usage	function	-
66	Sortimo International GmbH	automotive	structure, form, and function*	lightweight construction	-
67	Speedo Deutschland GmbH	textiles	structure, form, and function*	shark skin	market introduction / market penetration
68	Sto AG	construction	structure, form, and function*	lotus effect	market penetration
69	Suntec-Orga	energy	structure, form, and function*	"sunflower" solar collectors	market introduction

No.	Company	Sector	Biomimetics Classification	Biomimetics Principle	Stage of Development (year)
70	Tiger Coatings	nanotechnology	structure, form, and function*	lotus effect	-
71	Trolining GmbH	construction	structure, form, and function*	lotus effect, shark skin, form	market introduction
72	Velcro GmbH	textiles	structure, form, and function*	hook-and-loop closure function	market establishment
73	Villeroy & Boch	ceramics	structure, form, and function*	lotus effect	market introduction
74	VitrA Bad GmbH	sanitary wares	structure, form, and function*	lotus effect	-
75	Volvo	automotive	structure, form, and function*	lotus effect	market introduction

* Includes surfaces and interfaces
** Includes fluid dynamics
*** In order for the classification to remain clear, those products / processes which could not be assigned to a main category were designated as "extra."

[Source: own Internet enquiry]

4

Companies involved in biomimetics

The companies listed alphabetically in Table 12 show that the field of biomimetics exhibits an exceptionally broad spectrum, not only with respect to academic research, but also in the area of applications. To better assess the commercial biomimetics actors, the companies are classified according to industry. In order to allow for a certain degree of comparison, the industrial sectors are very broadly interpreted. Additionally, the products and processes (in some cases, also services) of the respective companies are assigned to a single class; as a result, the biomimetics classification is likewise rather broad and was based on Nachtigall (2002). Where the information was available, the biomimetic principle applicable to the respective products/processes (e.g., lotus effect) is also given in order to make quantitative clusters visible. Finally, the last column notes the current phase of the innovation process for each product or process in order to give a qualitative impression of the progress and successes in biomimetic innovation. Here a distinction was made between the three (chronologically) ideal phases of innovation: prototype/demonstration facility, market introduction, and market penetration.

Concrete biomimetics-related activities of enterprises

Again it must be assumed that a number of companies are conducting biomimetic research and development and/or marketing biomimetics-related products without doing so explicitly. Some of the commercial firms named in interviews with partnering universities and/or non-university R&D facilities do not make their involvement public at least not via the

Internet. Okalux is a good example: the company manufactures and markets a thermally insulating transparent glass for windows and facades that was based on the model of polar bear fur, but makes no public connection whatsoever between their work and biomimetics.[76] The results of our investigation of commercial firms also reflect a high degree of uncertainty, particularly with respect to the intensity and quality of the biomimetics involvement.[77]

Lotus effect dominating

Striking in the compilation of firms and the products and processes that they offer is the very clear dominance of the so-called *lotus effect*, in which the surface structure of the lotus leaf is technically emulated in order to achieve specific surface properties such as "self-cleaning." Almost half of all products and processes found in Table 12 can be assigned to this area; however, in at least some cases doubt exists about the extent to which a transfer of biological knowledge into a technical product actually occurred, or whether the label "lotus" is really being unjustifiably used for marketing purposes only. In such cases, only by specifically making a physical examination of the respective product would it be possible – if at all – to validate the claim, as the company involved would most likely not willingly admit to a misuse of the (relatively successful) lotus effect.

Much basic and applied research but only a few market-ready products and processes

Furthermore, in the investigation of the firms having a connection to biomimetics, it becomes clear that (in the case of products and processes for which information is available) only in a minority of the cases have the respective products or processes already been brought to market or established. The overwhelming number of cases involve prototypes or demonstration processes – about which, in many cases, no specific data can be obtained (trade secrets) – and it can be assumed that the products/processes are only in an early stage of R&D. Thus we arrive at a picture of the German biomimetics innovation landscape in which few successful innovations are widely established, some innovations are to be expected short- to mid-term, and much (if not the greatest share) remains in the dark and resists a detailed analysis.

4.4 Networking activities and intensity among biomimetics actors

Relations between biomimetics actors and BioKoN / KN-Biomimetics

In the course of identification, the impression initially arose that among the various actors – with the exception of the research networks BioKoN

76 See the documentation of the traveling exhibition "Inspiration Nature – the Bionics Patent Workshop" of the Deutsche Bundesstiftung Umwelt (DBU), 18–19; Download under: http://www.dbu.de/phpTemplates/publikationen/pdf/111206120202e29f.pdf (accessed: January 2007).

77 The empty cells in Table 12 indicate those situations where, even after relatively intensive inquiry, the missing information could not be obtained.

and KN Biomimetik – there would be relatively few connections. Therefore, in the next step we looked for possible connections, if any, among the various biomimetics actors. In this context, *connection between two or more actors* means a documented relationship between the actors – either simply a contact, or else a cooperative agreement or institutional relationship (for example, official involvement in a network). The results of this investigation are presented graphically in Figure 5 and Figure 6.

Networking of biomimetics actors

Figure 5 makes clear, first of all, that the two biomimetics networks appear larger when not only those actors are included that are registered as members, but also those to whom a mere connection exists (in green, BioKoN; blue, KN Biomimetics; and turquoise, both). At the same time, the chart shows that in many cases, particularly with the companies identified, no connection whatsoever to one of the research networks can be found (the grey areas). Certainly, there is always a degree of uncertainty as to whether no such cooperative relationships among the actors within the networks actually exist or whether the firms simply do not make this information public (for reasons of market strategy). However, the evidence seems to suggest that biomimetics in Germany still is being extensively undertaken outside of the existing institutional research networks.

Central role of BioKoN

In Figure 6, the (non-)existing connections between the actors identified are depicted in a somewhat different form. Again both networks are the focus of our considerations. In this chart, however, not only the connections to the networks are depicted, but also the relationships of the individual actors to one another. Thus, a total of three categories of intensity in the relationship between the actors could be distinguished, and these are represented by three different line weights in Figure 6: *contact* (narrowest line weight), *cooperation* (middle-weight), and *membership* (thickest line weight). The category "contact" would include, for example, an institution that organizes a conference on biomimetics and invites a representative of a company doing biomimetics research to speak. The category of *membership* includes, for example, the BioKoN and KN Biomimetik sponsoring institutions.

In Figure 6, the central role that BioKoN fills can clearly be seen. The role of KN Biomimetik as an important sub-center of biomimetics (*in the narrow sense*) in Germany is also recognizable. Furthermore, the illustrations demonstrate that although a great many actors maintain close relationships to one of the two research networks, among themselves they are only very loosely linked to one another. Many of the actors identified show no relationship whatsoever to other actors and are only linked to BioKoN or KN Biomimetics. However, there is also a series of actors that exhibit in part four or more connections to other biomimetics actors. These include, in particular, the universities in Saarbrücken, Freiburg, Tü-

Figure 5 — Memberships of universities, non-university R&D institutions, and companies in the BioKoN and KN Biomimetik research networks

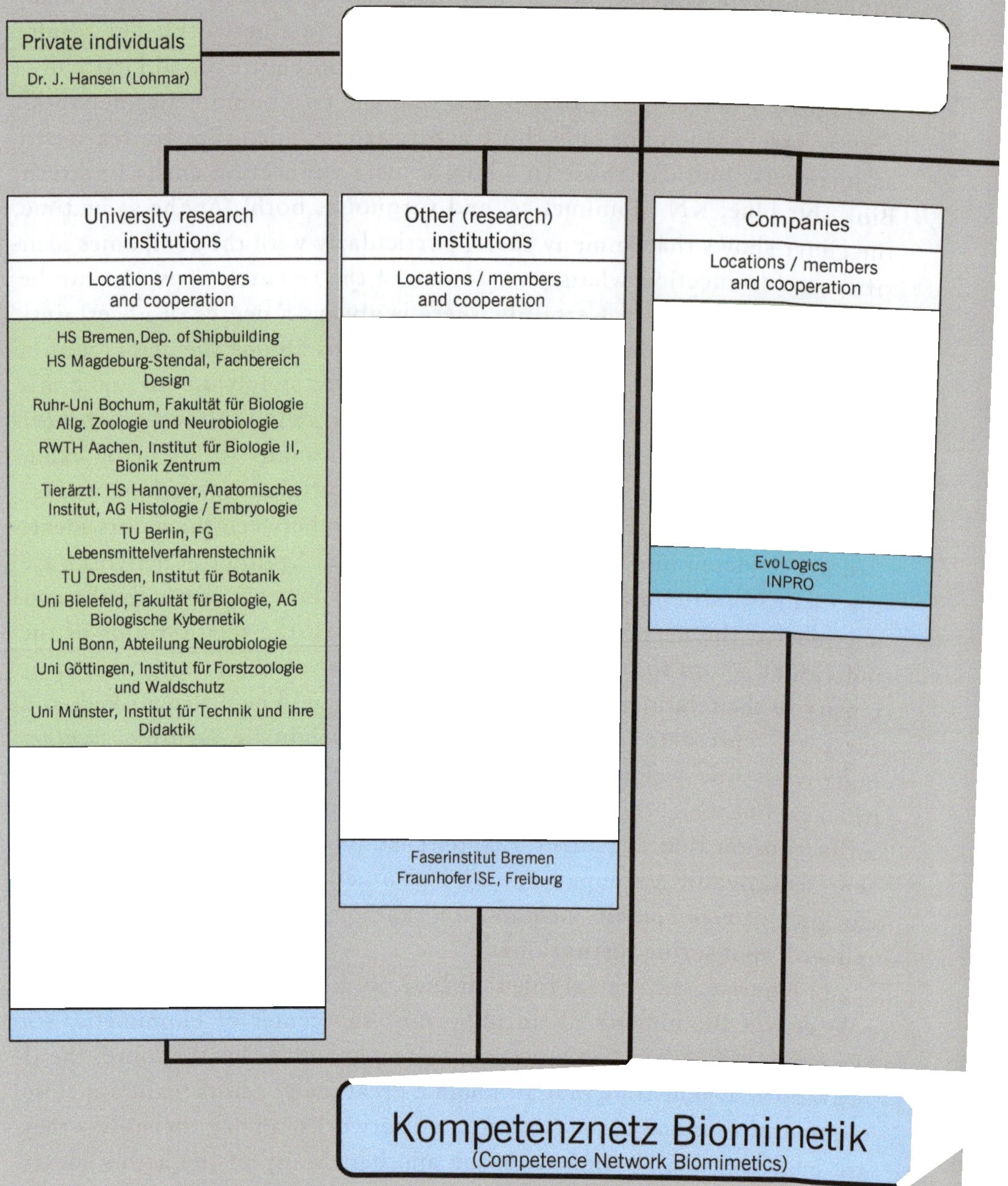

[Source: authors' own diagram]

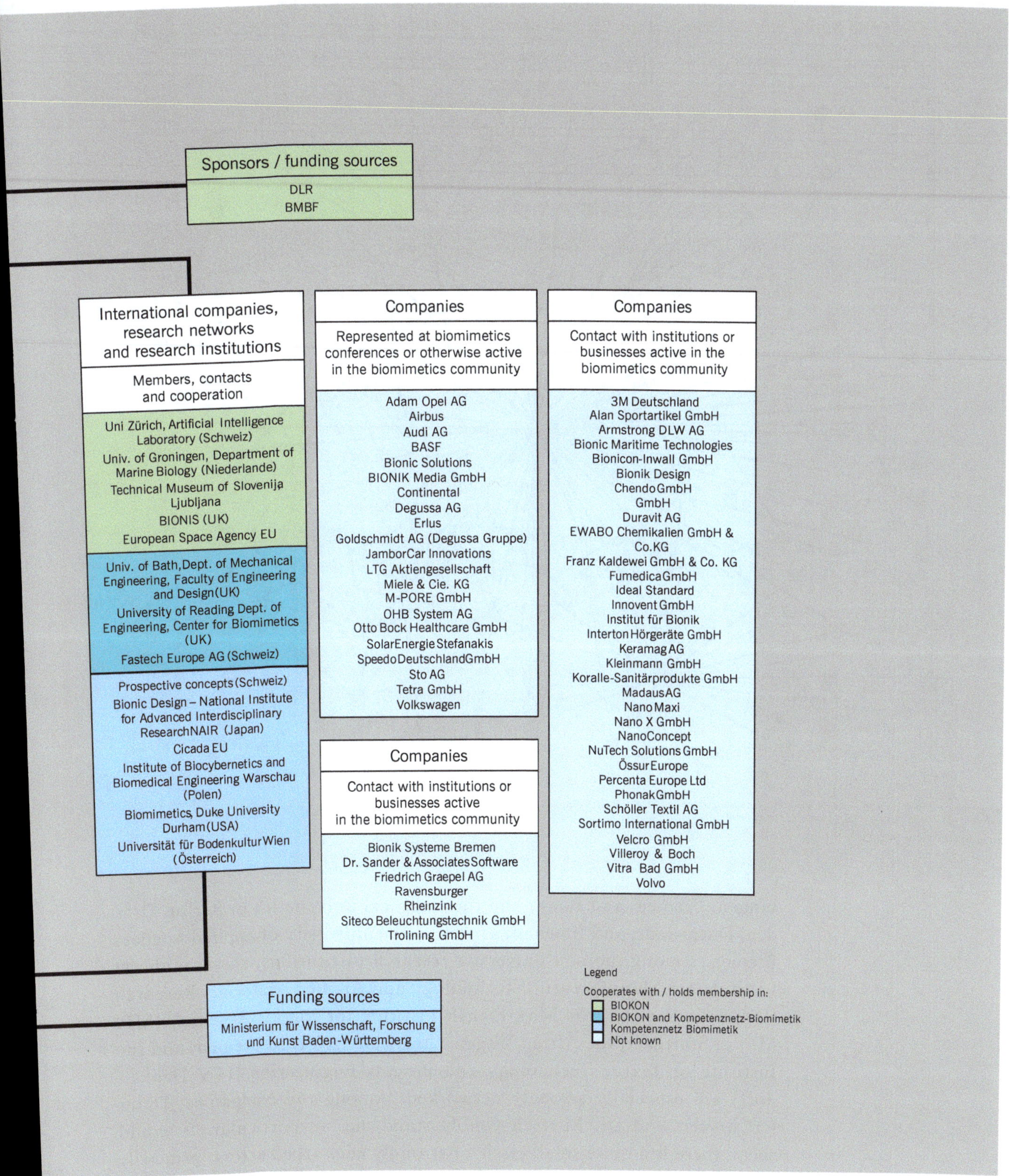
Sponsors / funding sources
DLR
BMBF
International companies, research networks and research institutions
Members, contacts and cooperation
Uni Zürich, Artificial Intelligence Laboratory (Schweiz)
Univ. of Groningen, Department of Marine Biology (Niederlande)
Technical Museum of Slovenija Ljubljana
BIONIS (UK)
European Space Agency EU
Univ. of Bath, Dept. of Mechanical Engineering, Faculty of Engineering and Design (UK)
University of Reading Dept. of Engineering, Center for Biomimetics (UK)
Fastech Europe AG (Schweiz)
Prospective concepts (Schweiz)
Bionic Design – National Institute for Advanced Interdisciplinary Research NAIR (Japan)
Cicada EU
Institute of Biocybernetics and Biomedical Engineering Warschau (Polen)
Biomimetics, Duke University Durham (USA)
Universität für Bodenkultur Wien (Österreich)
Companies
Represented at biomimetics conferences or otherwise active in the biomimetics community
Adam Opel AG
Airbus
Audi AG
BASF
Bionic Solutions
BIONIK Media GmbH
Continental
Degussa AG
Erlus
Goldschmidt AG (Degussa Gruppe)
JamborCar Innovations
LTG Aktiengesellschaft
Miele & Cie. KG
M-PORE GmbH
OHB System AG
Otto Bock Healthcare GmbH
SolarEnergie Stefanakis
Speedo Deutschland GmbH
Sto AG
Tetra GmbH
Volkswagen
Companies
Contact with institutions or businesses active in the biomimetics community
Bionik Systeme Bremen
Dr. Sander & Associates Software
Friedrich Graepel AG
Ravensburger
Rheinzink
Siteco Beleuchtungstechnik GmbH
Trolining GmbH
Companies
Contact with institutions or businesses active in the biomimetics community
3M Deutschland
Alan Sportartikel GmbH
Armstrong DLW AG
Bionic Maritime Technologies
Bionicon-Inwall GmbH
Bionik Design
Chendo GmbH
GmbH
Duravit AG
EWABO Chemikalien GmbH & Co.KG
Franz Kaldewei GmbH & Co. KG
Fumedica GmbH
Ideal Standard
Innovent GmbH
Institut für Bionik
Interton Hörgeräte GmbH
Keramag AG
Kleinmann GmbH
Koralle-Sanitärprodukte GmbH
Madaus AG
Nano Maxi
Nano X GmbH
NanoConcept
NuTech Solutions GmbH
Össur Europe
Percenta Europe Ltd
Phonak GmbH
Schöller Textil AG
Sortimo International GmbH
Velcro GmbH
Villeroy & Boch
Vitra Bad GmbH
Volvo
Funding sources
Ministerium für Wissenschaft, Forschung und Kunst Baden-Württemberg
Legend
Cooperates with / holds membership in:
BIOKON
BIOKON and Kompetenznetz-Biomimetik
Kompetenznetz Biomimetik
Not known

Figure 6 — Type and degree of networking among the individual universities, non-university R&D institutions, and companies

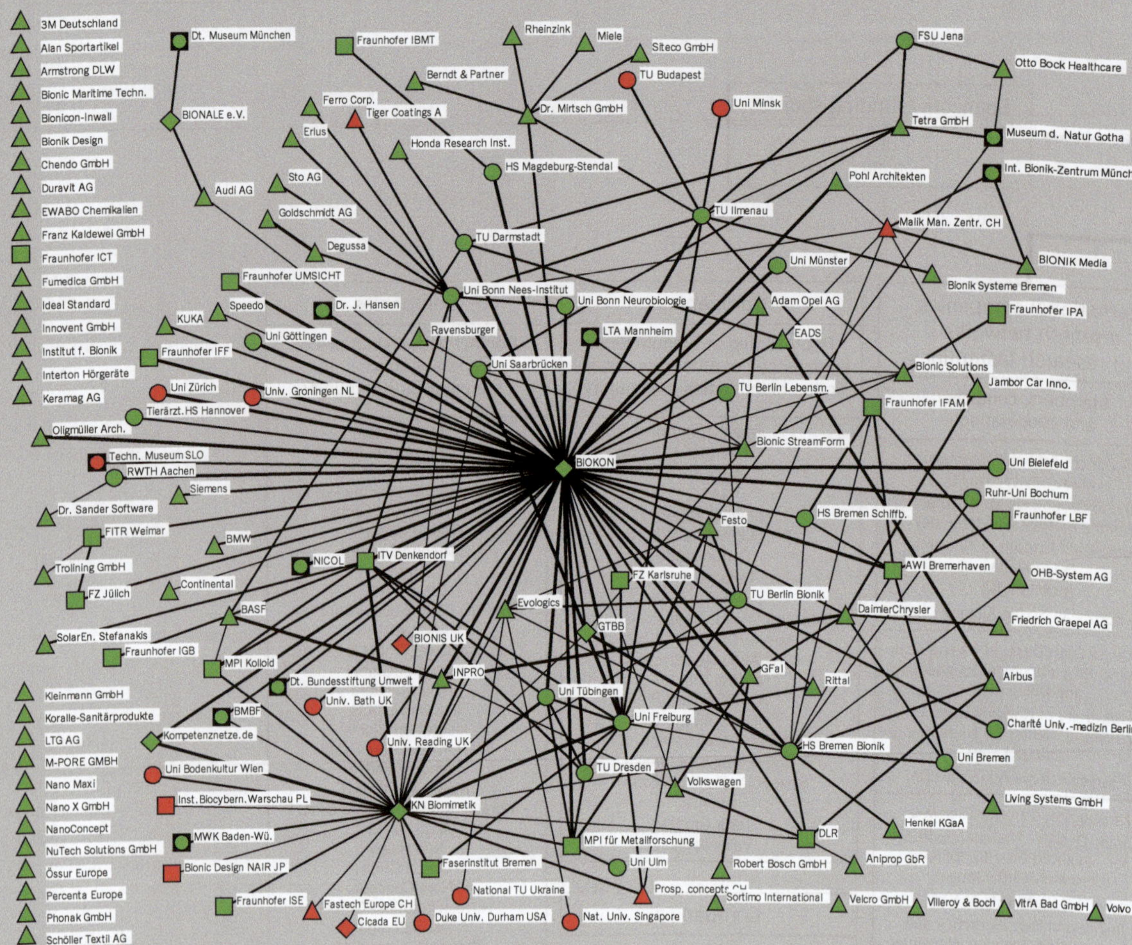

Legend: circle = university, square = non-university R&D, triangle = company, diamond = network node, blue = German, red = non-German; the thickness of the line is proportional to the degree of networking

[Source: authors' own diagram]

bingen, Bremen, and Bonn, and the technical universities in Berlin, Dresden, Darmstadt, and Ilmenau, as well as the university of applied sciences Bremen. Among the non-university research institutions, the Fraunhofer Institute for Manufacturing Technology and Applied Materials Research (IFAM) in Bremen, the Max Planck Institute for Metals Research (MPI MF) in Stuttgart, the Alfred Wegener Institute (AWI) in Bremen, and the Institute of Textile Technology and Process Engineering (ITV Denkendorf) are especially strongly networked. Among the companies, DaimlerChrysler and Dr. Mirtsch GmbH stand out, in particular. It would seem, there is a quite small circle of strongly networked actors, primarily dominated by the universities, and a large circle of actors whose only connection is with BioKoN or KN Biomimetik. Finally, it is clear that a whole series of establishments – almost all companies – are not (visibly) networked in any way whatsoever. These are represented on the lower and left sides of Figure 6.

Altogether, a review of the links between the actors identified shows that biomimetics (R&D) in Germany is presently only to a lesser extent networked. For the most part, biomimetics is being carried out by the various actors independently of each other.

4

4.5 Analysis and description of the German R&D landscape on the basis of literature reviews

Analysis of literature databases

Along with the approach described in the preceding chapter, in the course of the study a second, more fundamental approach to analyzing the German R&D landscape with respect to biomimetics was followed. This essentially involved searching for specific, biomimetics related terms in one of the largest and – with respect to its significance for the international scientific community – most important databases of scientific articles. In a first approach we restricted our search efforts to the geographical area of Germany and the time period 1990 through 2005. The results of the article search were then listed and sorted according to the particular institution that had published the respective articles (e.g., according to frequency of publication). The result was a picture showing how actively German institutions are publishing with respect to specific biomimetics terms. Under the assumption that the intensity of publishing is more or less proportional to the actual research and development activities, an impression of the German biomimetics R&D landscape was thus obtained. Finally, the search hits in the article database have been displayed in such a way that it became clear how often any one institution had published jointly with another institution. Following the assumption that joint R&D activities are more or less proportional to the frequency of joint publishing activities, one gets an impression of whether and how closely the various biomimetics players in Germany are linked together.

Web-of-Science databases

The article databases in which the literature search on German biomimetics R&D was carried out comprise the Web of Science Database (WSD) and are compiled and maintained by Thomson Scientific (previously the Institute for Scientific Information). The WSD consists of a total of five databases, in which separate searches are also possible. They comprise a total of approx. 8,800 scientific journals and periodicals and are updated on a weekly basis.[78] The authors of the databases attempt to compile an extensive but not necessarily all-encompassing database of scientific journal literature. They justify this with the argument that analyses of publishing frequency have shown that roughly 2,000 periodicals publish 95% of the articles cited.[79] As a result, the publishers of the WSD see their assignment in maintaining a database that documents the most essential research results.[80]

On the basis of the WSD, Thomson Scientific also prepares the so-called Journal Citation Reports. In these reports the periodicals are analyzed by citation frequency of their articles. A high frequency of citation for a journal thus increases its esteem in the scientific community, as a much cited publication is assumed to have a higher scientific relevance. The Citation Reports thus influence the behavior of researchers, who attempt to publish their articles in the most-cited periodicals. It can therefore safely be assumed that WSD is quite capable of representing the mainstream of research.

Search terms for *biomimetics in the narrow sense*

In order to represent the research landscape of *biomimetics in the narrow sense,* searches of the WSD using the terms *bionic** and *biomim** were performed.[81] Titles, keywords, and abstracts (when available) of the articles in all WSD databases were searched; the Boolean "OR" search operator was selected. (As a rule, authors use only one of the two terms – only rarely are both terms used in one and the same article.)

English and German articles

In theory, the choice of search terms thus also included the German term *Biomimetik* and its variations (*biomimetisch*, etc.). However, the WSD lists almost no entries in the German language, since the majority of internationally significant periodicals – and only these are represented in

78 The individual databases are:
- *Science Citation Index Expanded™*
  This database contains approx. 5,900 journals from 150 science disciplines. From 1991 onwards, the database also indexes abstracts for approx. 70% of all articles recorded in the database.
- *Social Sciences Citation Index®*
  Fully covers more than 1,700 journals from 50 social science disciplines. In addition, single entries from 3,300 natural science journals. Abstracts for approx. 60% of all articles in the database have also been indexed since 1992.
- *Arts & Humanities Citation Index®*
  Contains a total of 1,244 humanities journals.
- *Index Chemicus®*
  The database exclusively indexes chemical compounds.
- *Current Chemical Reactions®*
  Contains the documentation of chemical reactions.

The WSD allows analysis of indexed articles by authors, authors' addresses, language, year of publication, source, and discipline (among others).

79 Garfield, E., "The Significant Scientific Literature Appears in a Small Core of Journals." The Scientist V10 (17), 2 Sept. 1996.

80 See the Web site of Thomson Scientific under: http://scientific.thomson.com/free/essays/selectionofmaterial/journalselection/ (accessed: January 2007).

81 The asterisk '*', as used here in the search terms *bionic** and *biomim**, represents any number of following letters or numbers; i.e., with this search all articles can be found in which any one of the keywords *bionic, bionics, bionical*, or *bionically* appears (for the search term *bionic**) or *biomimetic, biomimetics, biomimetical*, or *biomimetically* (for *biomim**) .

the WSD – are English-language. Thus, effect of chosing *bionic** as the second search term,[82] resulting in the German term *Bionik* and its variations (*bionisch*, etc.) not being searched for, was negligible. A separate search for the term *Bionik* resulted in only one hit for the given time period and none for *bionisch**, making the authors believe that the selected combination of *bionic** and *biomim** has been reasonable and correct for the purposes at hand.

Search terms on »learning from nature«

In order to analyze the research landscape for *learning from nature,* a search in the WSD which has been carried out, in addition to the terms *bionic** and *biomim**, also included the alternative terms *self-repair**, *self-heal**, *biolog* inspir**, *bioinspir**, and *self organi**, all likewise associated with biomimetics.[83] All other parameters, as well as the procedures, were carried out in exactly the same way as was done in the search for *biomimetics in the narrow sense.*

4

Biomimetics in the narrow sense

In the analysis of the research landscape looking for *biomimetics in the narrow sense,* articles have been identified whose authors explicitly categorize their work under one of the two terms. In the course of the case studies (see chapter 3.2), as well as through the expert interviews, it became obvious that in many cases R&D work is being done which could very well be described as biomimetic but without the scientist necessarily ever making use of the terms *biomimetics* or *bionics*. Instead, alternative terms are being used that nonetheless have at their core a thoroughly biomimetic meaning (for example "self-healing") and that in essence describe biomimetic approaches and concepts. Therefore, in addition to a literature search on *biomimetics in the narrow sense,* a further enquiry using alternative search terms having a reference to *learning from nature* was performed.[84] The results of this search procedure are presented in the following section.

See chapter 3.2

Use of the terms »bionics« and »biomimetics«

Figure 7 and Figure 8 show that for the time period under investigation the number of publications in the WSD which use the terms *bionic** or *biomim** and originate in Germany greatly increased – not only in absolute numbers, but also in relation to the total number of all articles originating in Germany. The share of articles published based on the narrow interpretation of biomimetics (described above) increased from 1990

82 Use of the search term configuration *bionic** instead of *bioni** was necessary because in a trial search using *bioni**, a preponderance of hits was returned for articles dealing with bio-nickel ("bio-Ni") and clearly having nothing to do with biomimetics.

83 It has already been noted several times that very frequently terms such as *intelligent* or *smart* are associated with biomimetics approaches; for obvious reasons, queries using the terms *intelligent* or *smart* do not return any useful results, and therefore could not be included.

84 See particularly the beginning of chapter 3 for our approach to a differentiation between the categories of *biomimetics in the narrow sense* and *learning from nature.*

Figure 7 — Number of publications originating in Germany using the terms bionic* or biomim* and number of all publications in Germany in the WSD from 1990–2005 *

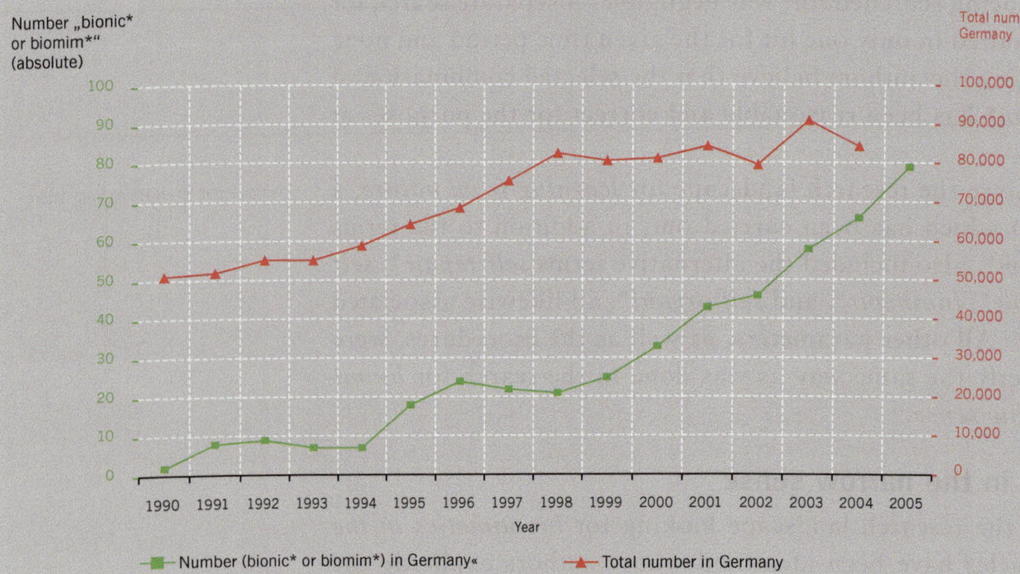

* For the series "all publications from Germany" no data could be made available, since in this year the number of publications exceeded 100,000 and could no longer be precisely indicated by the database

[Source: Compiled using data from the Web of Science databases.]

Large increase in publications on »bionics« and »biomimetics«

to 2004 by one order of magnitude, thus providing clear evidence for the strong growth in the number of biomimetics or bionics articles from 1999 onward.

The search for articles from German institutions using the terms bionic* or biomim* between 1990 and 2005 in the periodicals analyzed by the WSD resulted in a total of 468 hits.

Institutions related to »bionics« and »biomimetics« in Germany

Since a representation of all the institutions that were found as a result of the search described would be difficult to follow and of little use for our purposes (there are more than one hundred), in Figure 9 only the twenty-five most actively publishing institutions in Germany are represented (circles). The size of the individual circles is proportional to the sum of the published articles found for the time period (in four steps, beginning with 2–5 articles; institutions that have published only one article are not included). Leading, with a total of 25 publications, is the Max Planck Institute of Colloids and Interfaces in Golm (near Potsdam), followed by Universität Würzburg (23) and Universität Regensburg (21). (A full list of the complete names, abbreviated in the illustration for the sake of brevity and clarity, can be found in the appendix).

Cooperations

Since the intent in the preceding chapter was not only to identify and describe the biomimetics actors in Germany but also analyze their rela-

Figure 8 — Share of publications using the terms bionic* or biomim* vs. all publications originating in Germany in the WSD from 1990–2004

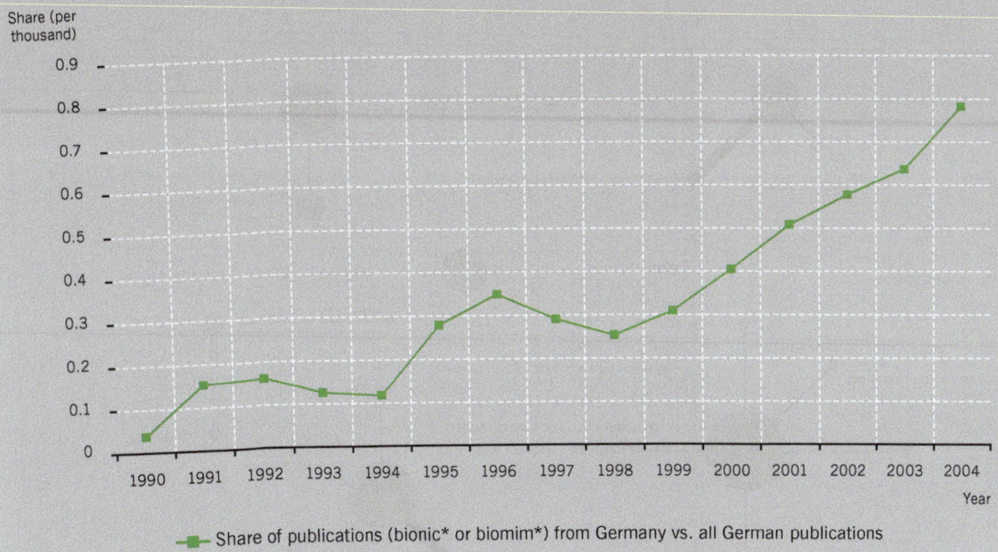

[Source: Compiled using data from the Web of Science databases.]

tionships to one another, in Figure 9 those institutions having jointly published with the twenty-five most published German institutions (under the terms *bionic** or *biomim** for the time period already mentioned) are also represented; this is indicated by a connecting line. Thus, German (blue square) as well as non-German (red triangle) institutions are represented. The line weight chosen proportionally reflects the frequency of the joint publishing activities, whereby, for reasons of readability and also meaningfulness, only those institutions were included that have had at least two joint publications.

Few joint publications

Figure 9 indicates that somewhat more than half (13) of the most actively publishing German facilities have not jointly published two or more articles together with another institution. Somewhat fewer than half of the institutions (11) have jointly published to some extent, with several institutions, two or more times. However, with these institutions, it also becomes clear that their publication activities take place for the most part independently and only rarely involve another institution. Interestingly, the preponderance of institutions appearing as co-authors is foreign. Finally, only three closer cooperations stand out: that between Universität Regensburg and Rice University in Houston, Texas (USA); between the Max Planck Institute for Polymer Research and Universität Mainz; and between Universität Ulm and Lomonosov Moscow State University (Russia).

Figure 9 — The Twenty-five German institutions most actively publishing using the terms bionic* or biomim* including local and foreign institutions with which articles using these terms were published; publication time period: 1990–2005 (further details in text)

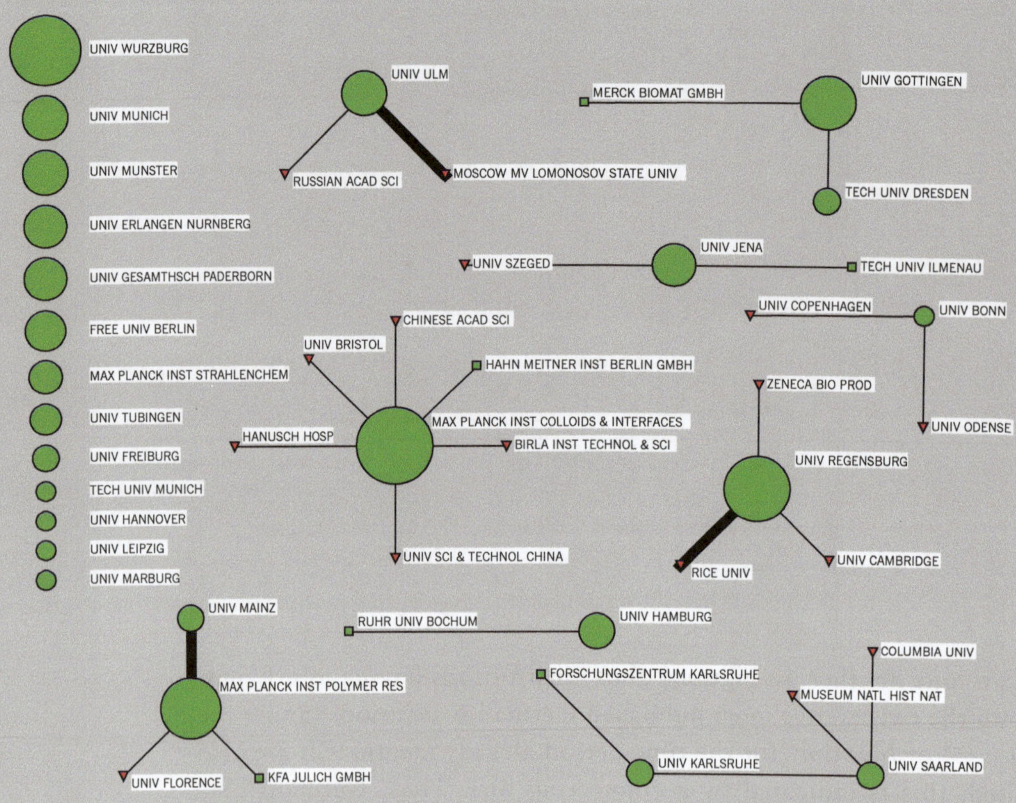

[Source: Compiled using data from the Web of Science databases.]

Discrepancy between literature research and network analysis

Figure 9 further makes clear that the make-up of the R&D actors, particularly among German universities and non-university research institutions as described earlier in chapter three, varies in part from the impression gained through the literature queries. None of the universities (for example, RWTH Aachen, TU Darmstadt, TU Berlin, Freiburg, Tübingen, Saarbrücken, Bremen, and the Hochschule Bremen) strongly represented in the two research networks, BioKoN and KN Biomimetik, and named by the experts interviewed as important biomimetics actors in Germany play a role – at least no outstanding role – according to publications in internationally prestigious professional journals. The situation with non-university R&D institutions is similar: only the Max Planck Institute of Colloids and Interfaces could be identified by means of the networks, expert interviews, conferences, and literature searches as being a significant biomimetics actor. There seems to be an entire series of actors in the area of German biomimetics R&D having little or no contact with the established network structures and at the same time cooperating only rarely with other actors, if at all, with respect to biomimetics activities. A spot check – on account of the many hundred publications, a detailed investigation would not be possible here – of these institutions showed that in the literature searches in the WSD, those institutions in particular stand out that are active in the areas of chemistry, biochemistry, biotechnology, and nanotechnology. It is exactly these areas that tend to be underrepresented in the biomimetics networks BioKoN and KN Biomimetik; this could explain the partial discrepancy between the results from the searches detailed in Sections 4.2 to 4.4 and the literature searches described here.

Learning from nature

As with the literature search queries for biomimetics in the narrow sense, in the course of the investigations of the biomimetics R&D landscape in Germany we also conducted an extensive search for published articles related to *learning from nature*.

Publications on »learning from nature« have increased strongly

As Figure 10 makes clear, the number of publications from Germany for the alternative biomimetics terms mentioned previously that we located in the WSD for the time period 1990 to 2004 increased greatly, and not only in absolute numbers but also relative to the number of all publications from Germany in the WSD for this time period. Noteworthy is that the number of publications that reflects a broad interpretation of biomimetics as a whole was approximately five times as large as the number of publications for biomimetics in the narrow sense. For the time period 1990–2005, a total of 3,091 publications for all the biomimetics terms mentioned previously were found in the WSD, of which only 468 could be allocated to the terms *bionic** and *biomim**. Thus it becomes clear that the biomimetics R&D approach, namely, the transfer of biological knowledge into non-biological (mostly engineering) contexts is more widespread than the use of the terms *biomimetic and bionic.*

Institutions related to »learning from nature« in Germany

With the graphical representation of the institutions from Germany publishing on biomimetics in the narrow sense in Figures 9 and 11, we present the twenty-five institutions in Germany found in the WSD for the time period 1990 to 2005 that have published most under the alternative biomimetics terms. Again, those institutions domestic and foreign, with which the previously named twenty-five top institutions have jointly published under the alternative terms are also represented (the graphical depiction is otherwise the same as in Figure 9).

Importance of the term self-organisation

Figure 11 shows yet a different picture of the German R&D landscape with respect to biomimetics than the previous diagrams and information. The large hub with TU Berlin at the center in the diagram stands out in particular. The publications of TU Berlin were therefore given a closer

Figure 10 — Share of publications originating in Germany using the terms bionic* or biomim* vs. all publications originating in Germany in the WSD, 1990–2004 (further details in text)

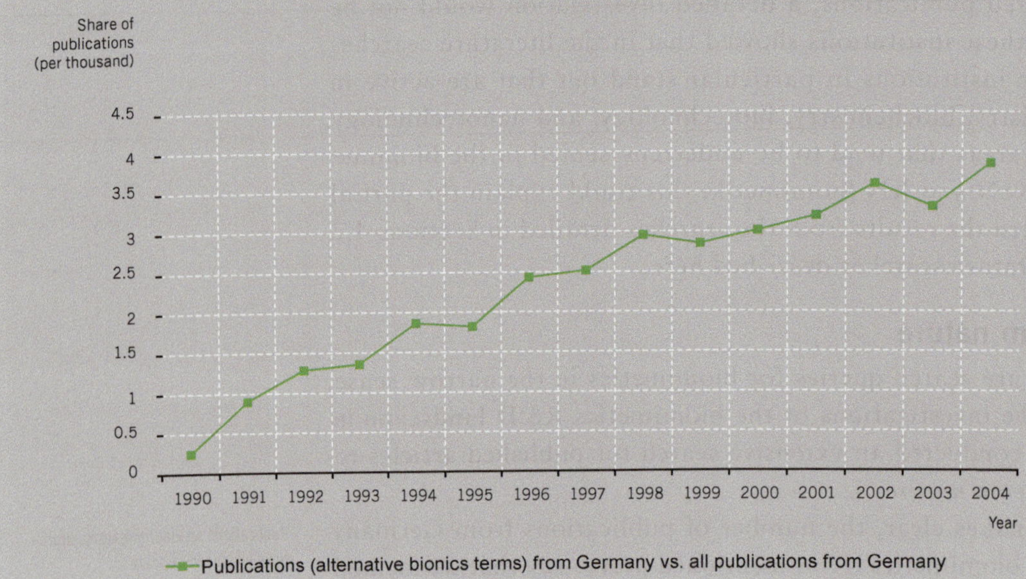

[Source: Compiled using data from the Web of Science databases.]

look, revealing that 217 of the total 227 publications fell under the term *self organi**. Of these, 172 can be assigned to physics in the broadest sense, i.e. inclusive of physical chemistry. An inquiry to determine whether any one particular institution of the university is especially active revealed that at least 168 of the publications originated at the Institut für Festkörperphysik (Institute for Solid-State Physics) and/or with the participation of the institute director, Professor Bimberg. The institute is intensively researching nanomaterials and nanophysics, and the processes and structures of self-organisation that play a large role. Here again, we find areas only barely or not at all covered by the biomimetics networks BioKoN and KN Biomimetik that are very strong in publications. It is interesting, with respect to the central position of TU Berlin, that among the 227 hits for publications in the WSD scarcely any from Prof. Rechenberg's institute are found – even though it plays a significant role in the research networks.

Few joint publications

As in Figure 9, in Figure 11 we see that roughly half of the institutions identified that published under the alternative biomimetic terms did so exclusively independently, i.e. without cooperating with other institutions. Those institutions that did publish jointly in most cases did it with only a single partner institution and these cooperative efforts resulted in relatively few articles. The degree of networking with respect to publications for biomimetics in the broad sense *(learning from nature)* can therefore be judged as quite small.

Figure 11 — The twenty-five German institutions most actively publishing using alternative biomimetics terms including the domestic and foreign institutions where articles using these terms were published (publication time period: 1990–2005)

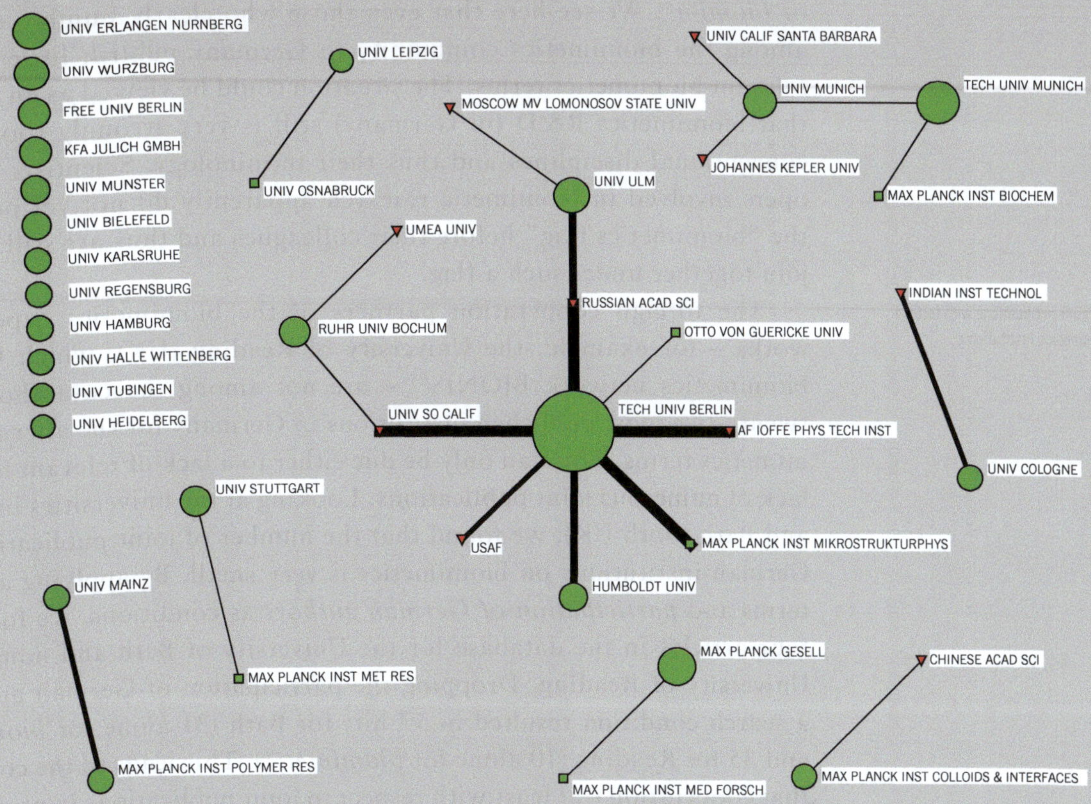

[Source: Compiled using data from the Web of Science databases.]

Discrepancy between literature research and network analysis

Furthermore, it generally was the case that while searching the 3,091 publications using the broad biomimetics terms, that those actors making up the core of the explicit biomimetics community (see chapter 4.2) were barely represented, or even not at all. Of those university and non-university institutions that are members of BioKoN, there are some that are not found in neither Figure 9 nor Figure 11, the majority of the others are somewhere in the middle. However, TU Berlin is the exception; as already noted, the large number of publications can be allocated to the Institute for Solid-State Physics and not to the "biomimeticists" in Prof. Rechenberg's insititute.

A specific search for known and extremely active members of the explicit German biomimetics community found in comparison only a few hits. The search terms *lotus* (topic) and *Barthlott* (author) led, for example, to (only) six hits; however none of these were picked up by the search for *bionic** or *biomim**. A further search for *Bannasch R* (author) resulted in ten hits that, in fact, do relate to the topic of biomimetics; however, only one of these articles can be identified using the search terms *bionic** or *biomim**. We see here that even those who clearly number themselves among the biomimetics community in Germany publish little under the relevant biomimetics terms. The situation could be viewed as an indication that biomimetics R&D (in Germany) still is very strongly rooted in the conventional disciplines and thus their terminology. Scientists and developers involved in biomimetic research apparently do not yet publicly fly the "biomimetics flag" before their colleagues and thus are still unable to join together under such a flag.

Few joint publications with foreign biomimetics institutions

The foreign cooperation partners of the biomimetics expertise networks – for example, the University of Reading (UK), which funds the biomimetics network BIONIS[85] – are not among the co-authors of the twenty-five most published institutions in Germany for the alternative biomimetics terms. This can only be due either to a lack of relevant terms or a lack of numerous joint publications. Looking at the universities in Reading and Bath (both UK), we found that the number of joint publications with German institutions on biomimetics is very small. By applying all search terms and *participation of German authors* as conditions, we found only four articles in the database for the University of Bath and none for the University of Reading. Dropping the participation of German authors as a search condition resulted in 57 hits for Bath (31 alone for *biomimeti**) and 35 for Reading (10 alone for *biomimeti**). This leads to the conclusion that cooperation – at least with respect to joint publications in professional journals of international renown – between the German and the British biomimetics community is not particularly extensive.

Altogether, these investigations suggest that in addition to the highly visible public actors in the German biomimetics community, there are apparently still other numerous institutions doing research on biomimetics or related topics, but not under the biomimetics or bionics label.

4.6 International Biomimetics R&D landscape

For the analysis of the international R&D landscape with respect to *biomimetics in the narrow sense* and the approach *learning from nature*,[86] we expanded our analysis of publications in the Web of Science to include a

85 See the Web site of the BIONIS network under http://www.extra.rdg.ac.uk/eng/BIONIS/ (accessed: December 2006)

86 See particularly the beginning of chapter 4.1 for our approach to a differentiation between the categories of *biomimetics in the narrow sense* and *learning from nature*.

patent analysis.[87] The latter was performed by the patent information center PATON at Universität Ilmenau. For the sake of clarity, an in-depth examination was limited to the USA, Japan, Germany, France, England and China.

Patent and literature analyses of *biomimetics in the narrow sense* and *learning from nature*

Research in »web-of-science« databases

Once again, we first searched the Web of Science for the time period 1990–2005 using the terms *bionic** and *biomim** (biomimetics in the narrow sense) and then with the terms *biolog*-inspir*, bioinspir*, self-organi*, self-heal**, and *self-repair** in the same time period (learning from nature).[88] The total number of professional articles for *bioni** and *biomim** in the time period under investigation also increased steadily internationally (see Figure 12).

A total of 4,878 publications were recorded worldwide. Of these, 3,728, or roughly 76%, occurred in the countries slated for in-depth investigation. The largest number, by far, were published in the USA, with 1,586 articles. Japan (593), Germany (467), China (370), Great Britain (357) and France (355) were likewise among the top ten. The following illustration shows the distribution by country.

USA top of the list in literature releases

The dynamic of publications under the topic of *learning from nature* for the time period investigated was fully comparable with that of those under the narrow interpretation of biomimetics (see Figure 14). The annual number of publications has increased more than tenfold since 1990, but there are small differences in the distribution by country. Germany moves up to second place, and China drops still farther down. A total of 23,535 publications were recorded; of them 17,024 were from six countries: the USA (6,547 or 27.82%), Germany (3,089 or 13.13%), Japan (2,642 or 11.23%), France (1,776 or 7.55%), Great Britain (1,673 or 7.11%) and China (1,297 or 5.51%). The three countries with the most publications (USA, Germany, and Japan) were responsible for more than half of all the published articles worldwide for the expanded search terms.

Publications by the countries USA, China, France, England, Japan and Germany

Results of patent analyses

Parallel to the literature searches in the Web of Science database, patent analyses were also carried out as part of our investigation of the international biomimetics R&D landscape. The patent analyses were conducted for the time period 1995 to present using the search terms *bionic*, biomim**, and *bio(-) inspir** and show a trend similar to that of the results of the literature searches (see Figure 15).[89] A total of 504 patent publications

87 See section 4.5 for a brief explanation of the Web of Science databases.

88 See section 4.5 for more on the methods and the databases.

89 Databases searched: WPINDEX/WPIFV (42 most important industrial nations), Japio (Japan), IFIPAT (USA), EPfull (European Patent Office), PCTfull (World Intellectual Property Organisation), PATDPAfull (Germany). SPATFULL (USA). For the patent analysis we expanded the narrow interpretation of biomimetics to include the term *bio-inspired*. However, according to PATON, this addition only produced a marginal change in the number of hits as compared to searches for *bionic** or *biomimeti**.

Figure 12 — Worldwide development of annual publications on biomimetics in the narrow sense (bionic* and biomim*) in the Web of Science database

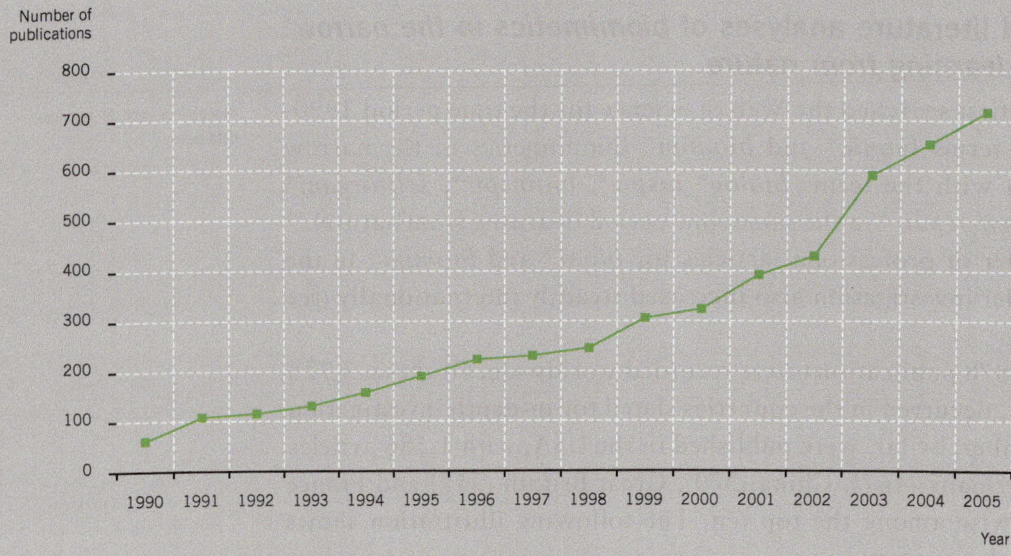

[Source: Compiled using data from the Web of Science databases.]

Figure 13 — Distribution by country of published articles in the literature for the time period 1995 to 2005 for *biomimetics in the narrow sense* (bionic* and biomim*)

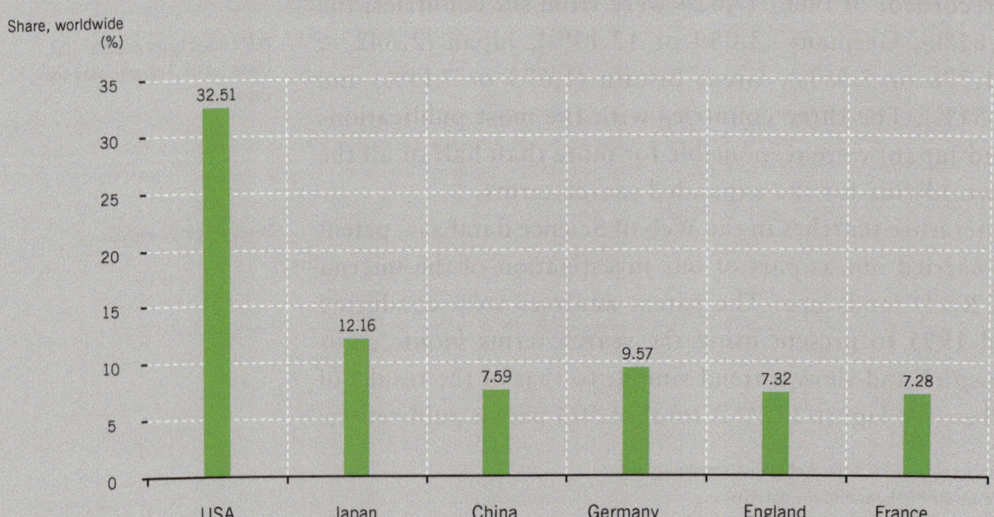

[Source: Compiled using data from the Web of Science databases.]

Figure 14 — Worldwide development of annual publications on the biomimetics terms (bionic*, biomim*, biolog*-insp*, bio(-)insp*, self-orga*, self-repair* and self-heal*) in the Web of Science database

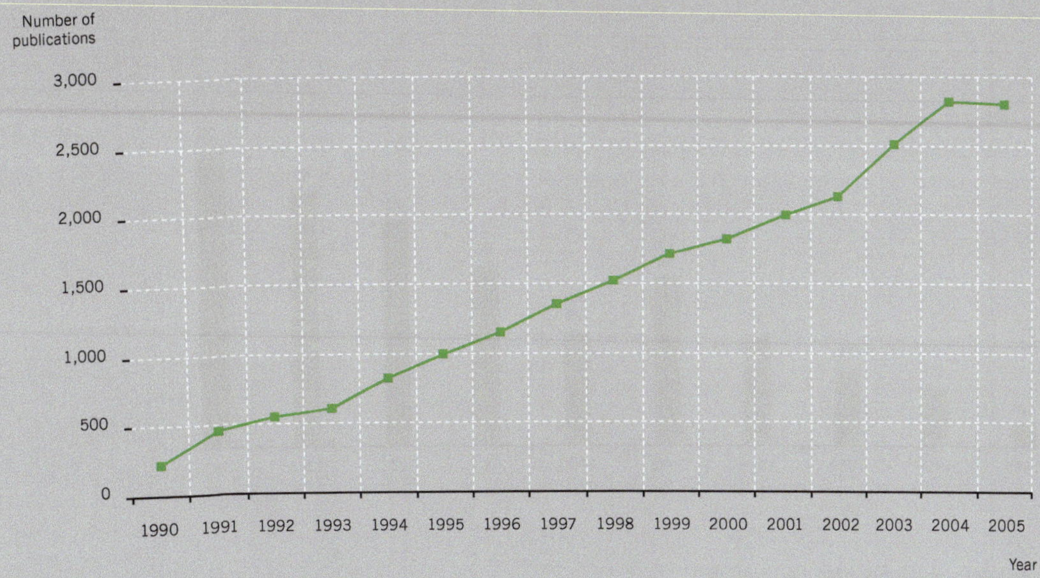

[Source: Compiled using data from the Web of Science databases.]

worldwide for the time period 1995 to 2004 were found that contained the terms *bionic**, *biomim**, and/or *bio(-)insp**; however, the number of annual publications increased more than tenfold from 8 in 1995, to 99 in 2004.

Increasing number of patents for »bionics« and »biomimetics« worldwide

The patent search results were similar for the extended biomimetics concept, *learning from nature,* (see Figure 16). In addition to the terms for biomimetics in the narrow sense, *self-organi**, *self-heal**, and *self-repair** were included. The number of hits for the time period 1995 to 2004 thus increased to 2,772, whereby the number of publications annually only increased by slightly less than a factor of five, from 104 (1995) to 481 (2004). However, the distribution of patent applications by country and region is significantly different from that for article publications (see Figure 17 in conjunction with Figure 13).

USA and Chinas top of the list

The most patent applications were recorded in China, closely followed by the USA. Japan and Germany form a second group, and Great Britain and France, a third.[90]

If one searches in the patents for the broad biomimetics interpretation (with the expanded search terms), the USA again takes a clear lead (1,055 patent applications), followed at a distance by Japan (764) and a greater distance China (470) and Germany (311) (see Figure 18).

90 The significance of the number of Chinese patents should not be overvalued. Sample analyses showed that patents registered for biomimetics in the narrow sense in China varied greatly from our understanding of biomimetics.

Figure 15 — Worldwide development of annual patent publications on biomimetics *in the narrow sense* (bionic*, biomim*, and bio(-)insp*)

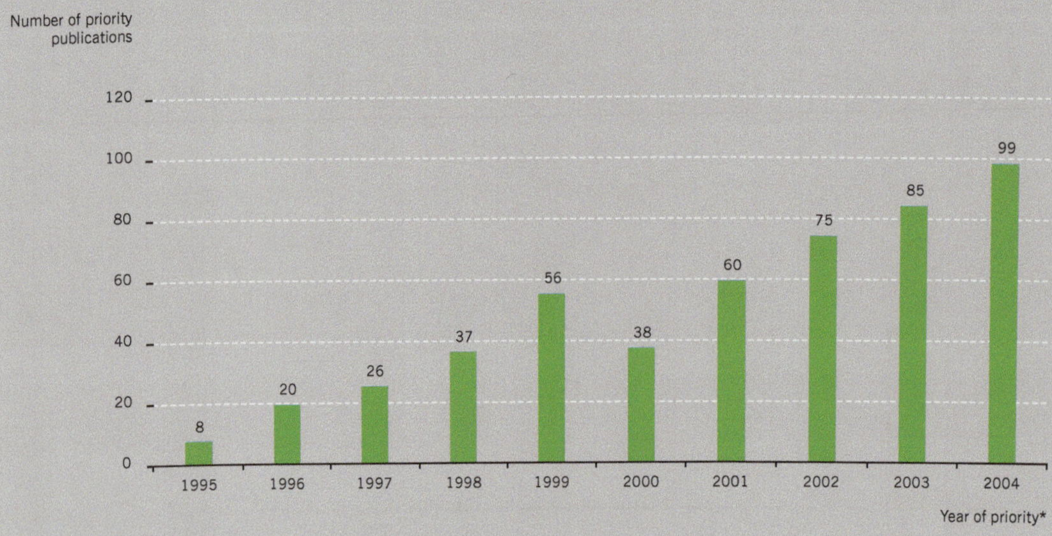

[Source: Paton Patentcenter, Thuringia GmbH, Worlddatabase WPIDS (Last Update 2007)
* About 18 Month between registration and publication]

Figure 16 — Worldwide development of annual patent publications on broad biomimetics terms (bionic*, biomim*, biolog*-insp*, bio(-)insp*, self-orga*, self-repair* and self-heal*)

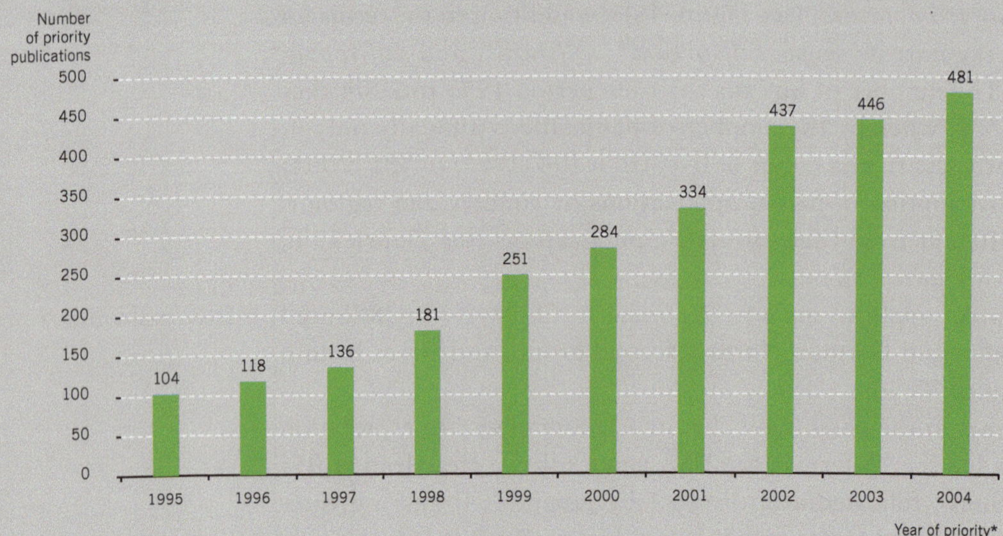

[Source: Paton Patentcenter, Thuringia GmbH, Worlddatabase WPIDS (Last Update 2007)
* About 18 Month between registration and publication]

Disciplinary distribution of publications

Figure 19 provides a further overview of the disciplines in the Web of Science database of publications for "biomimetics in the narrow sense" (*bionic** or *biomim**) for the time period 1995–2004; for the sake of clarity the twenty most frequent science disciplines were aggregated into broader categories (chemistry, physics, etc.). In the professional publications, a clear dominance can be seen in the areas of chemistry (physical, organic, multi-disciplinary chemistry). In each of the six of the countries investigated (China, France, Great Britain, Japan, the USA and Germany), more than half of the articles published fall into this area. The material sciences were also frequently mentioned (between 10% and 20%). However, these classifications should only be considered as a guide and not as evidence, since each of the individual articles in the Web of Science databases can be assigned to multiple, even diverse disciplines, a consequence is that it is not clear which patents have been assigned to which dicipline now, and how many disciplines. From the observation that more than 50% are assigned to chemistry, it is not possible to derive the conclusion that 50% of the patents originated in this discipline.

Classification of patent releases among R&D sections

In connection with the investigation of patents, in turn the research and development categories for which the patents were registered – alongside the scientific disciplines to which the patent applications originated – are particularly interesting. The analysis of the R&D categories for patents where the search terms *bionic** and *biomim** were used show here a picture similar to the one outlined for the literature analyses (see Figure 19). Chemistry and biochemistry are strongly represented as R&D categories the material sciences appear not to be present at all as a R&D category; the corresponding registrations are likely spread out over several categories (C08, C09, B29). A significant portion of the latter could possibly have been attributed (via prosthetics) to the category of medicine that thoroughly dominates here (see Figure 20).

It has already been mentioned several times that in the English language, the term *bionics* in particular, and its variants are closely associated with prosthetics and the replication of human organs and tissue. This could possibly explain the dominance of medicine here. But also in Germany, Japan and China, a large number of patents originate from the categories of medicine, chemistry and biochemistry, so that this cannot entirely explain the strong dominance in the category of "medical or veterinary science, hygiene." Here too, the individual patents can be assigned to several R&D categories, so that it is not possible to determine exactly how many patents are actually to be found behind each category.

Bionics and prosthetics / medicine

If one then takes the extended search terms *(learning from nature)*, as the basis the result is a very different picture. There is a lead group of five categories in which only the categories medicine and measuring and testing

Figure 17 — Patent publications by country for the time period 1995 to 2004 using the narrow biomimetics concept (*bionic and *biomimetic**)***

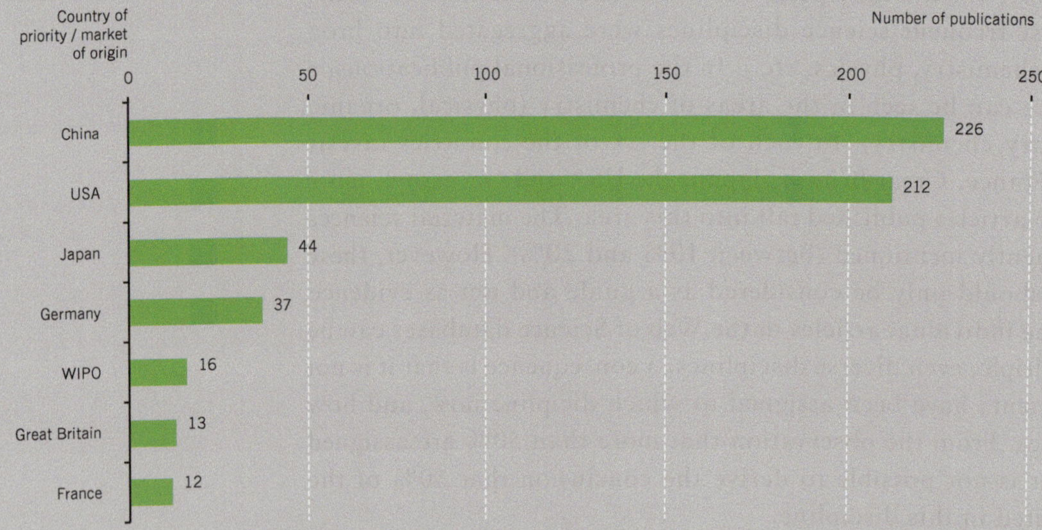

* "WIPO" refers to patents that have been registered worldwide (with the World Intellectual Property Organisation).
[Source: Paton Patentcenter, Thuringia GmbH, Worlddatabase WPIDS (Last Update 2007), - * About 18 Month between registration and publication]

Figure 18 — Patent publications by country for the time period 1995 to 2004 using the broad biomimetics concept (*bionic*, biomim*, bio(-)insp*, self-orga*, self-repair and *self-heal**)****

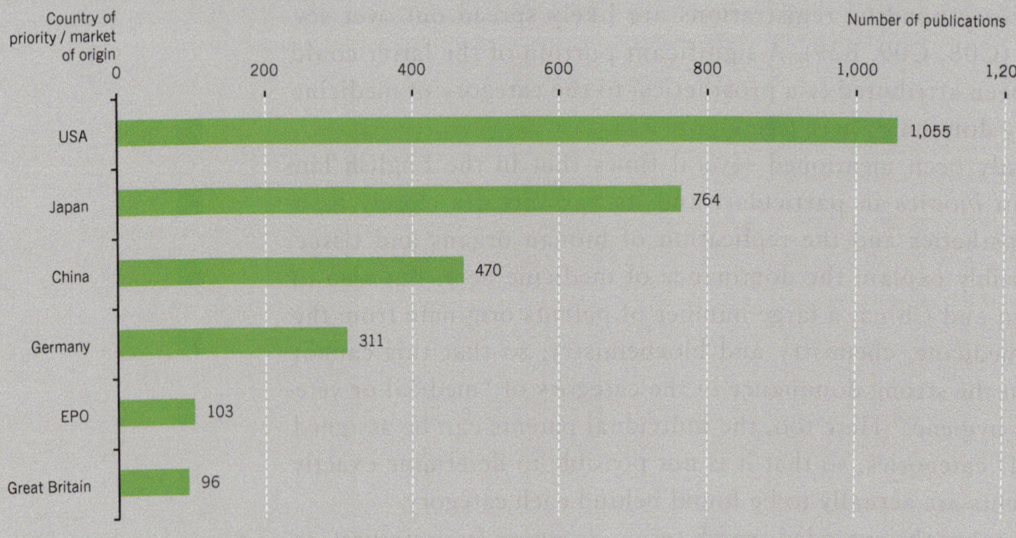

** "EPA" depicts the number of patents registered with the European Patent Office.
[Source: Paton Patentcenter, Thuringia GmbH, Worlddatabase WPIDS (Last Update 2007), - * About 18 Month between registration and publication]

Figure 19 — Literature publications by discipline and country for biomimetics in the narrow sense (bionic* and biomim*)

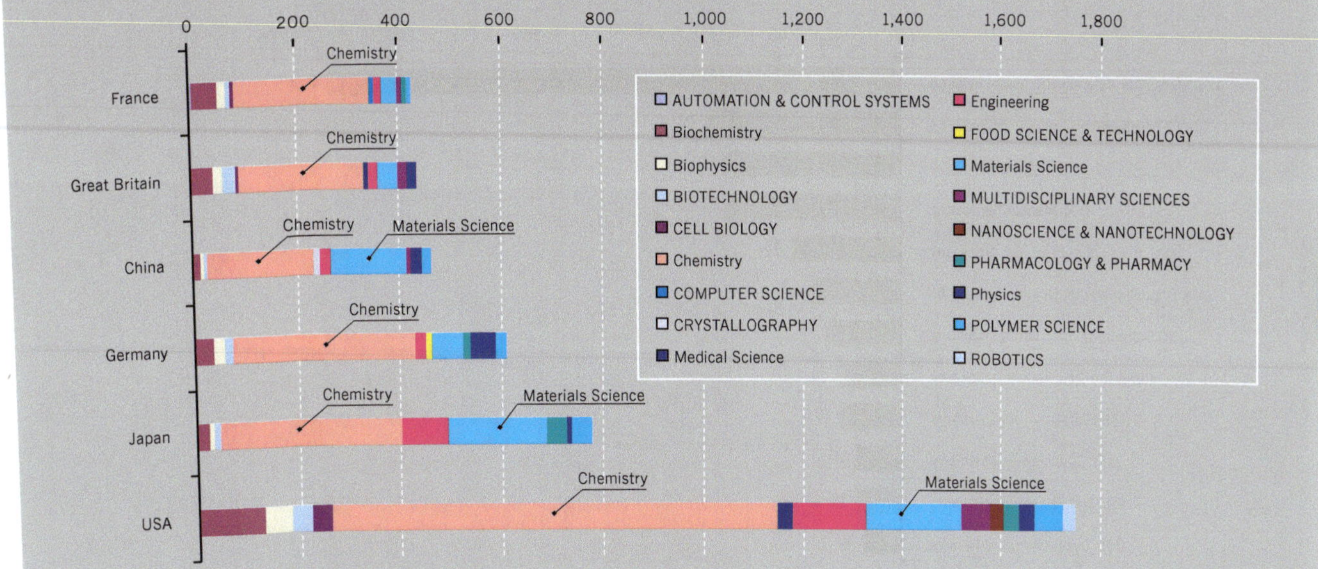

[Source: Compiled using data from the Web of Science databases.]

are identical with the lead group for queries using *bionic** and *biomim**. All other fields originate in the area of information and communication technology, and the category computing moves to the top of the list. Basic electrical elements and electrical communication technique then take the third and fourth place respectively. Only then, at a significant distance, do organic chemistry and biochemistry appear. The expanded search terms *self-organisation, self-healing,* and *self-repair* cover primarily additional ICT areas. But they also significantly increase the number of hits with material technology and nanotechnology topics.

Looking at the distribution by country of the patents for biomimetics in the narrow sense, we find that approximately half of the patent applications occur in the categories of medicine, biotechnology, and organic chemistry. In the USA and Japan noticeably more patents are recorded in the category measuring and testing. Germany shows a rather average distribution.[91]

National distribution of patents

The analysis of patent notifications for *learning from nature* reveals – as in the case of the literature analysis – a different picture. None of the countries shows a dominant discipline. In the USA, the categories of "computing, calculating, counting"; "electrical communication technique"; and

91 This statement is supported by more detailed analyses of the categories from which the patents originate; for the sake of brevity we depict only one graphic.

Figure 20 — Patent publications by R&D area for biomimetics in the narrow sense

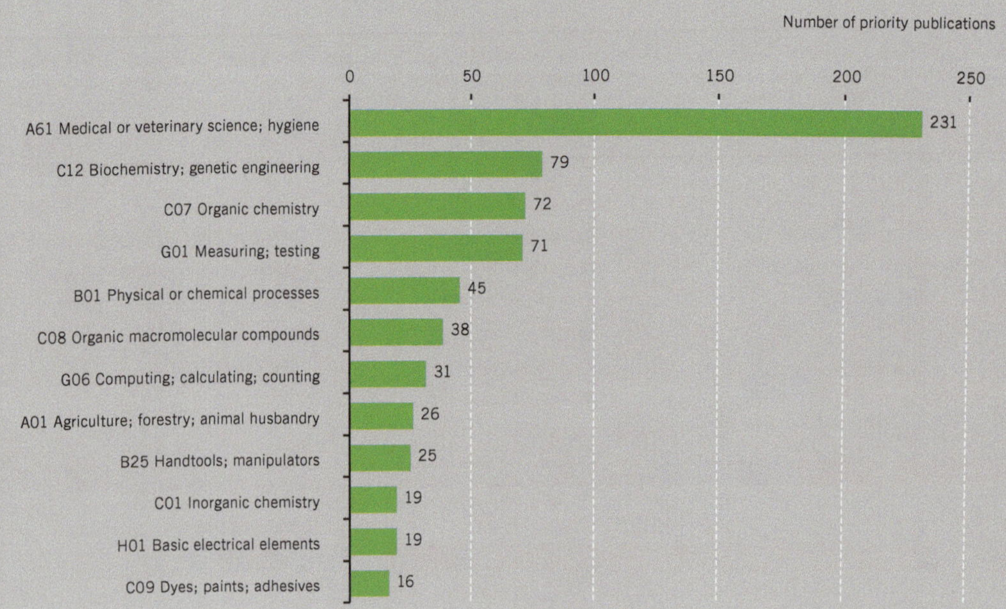

[Source: Paton Patentcenter, Thuringia GmbH, Worlddatabase WPIDS (Last Update 2007), - * About 18 Month between registration and publication]

Figure 21 — Patent publications by R&D area, search criteria expanded to include self-organi*, self-heal*, and self-repair*

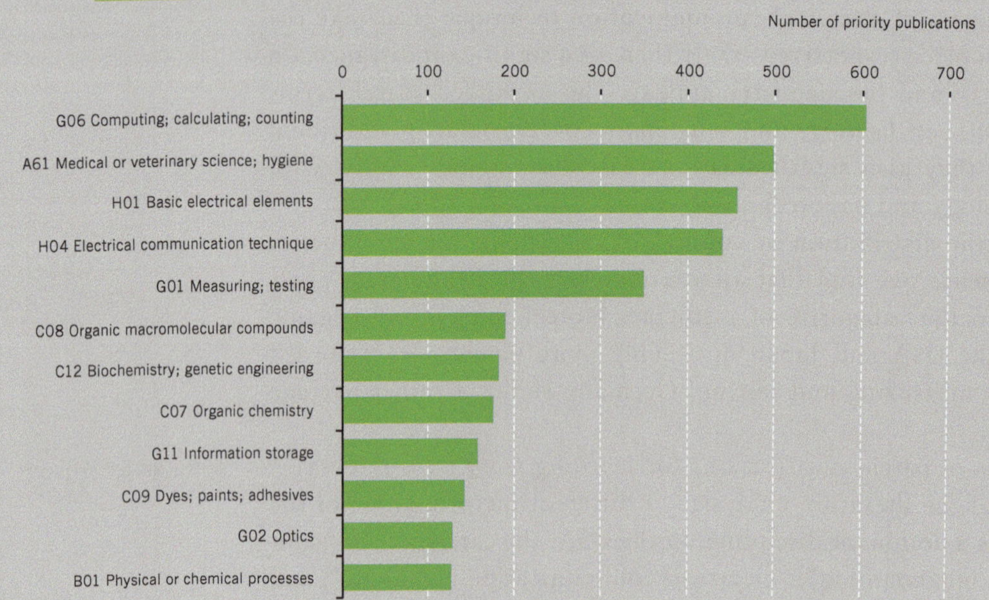

[Source: Paton Patentcenter, Thuringia GmbH, Worlddatabase WPIDS (Last Update 2007), - * About 18 Month between registration and publication]

"measuring and testing", in particular, appear to be important; in Japan, "basic electrical elements" and "computing, calculating, counting". In Germany, only the category "basic electrical elements" stands out somewhat.

If one concentrates on the differences between the narrow and the broad search criteria, it is striking that in the USA the share of patents in the areas "electrical communication technique", "measuring and testing", "information storage", and "organic chemistry" increase significantly if we use the expanded search terms as our basis. In Japan and Germany the share in the category "basic electrical elements" increases.

Most patents are applied for by businesses

An analysis of the actors shows that most of the patents that could be found using the expanded search terms were submitted by businesses.[92] The leaders, by far, are found in Japan; among them, Sony, Seiko, Toshiba, and Fuji; from Germany Infineon and Siemens could be counted among the top twenty with the most patent applications; from the USA, IBM, the University of California and Motorola; from Korea, Samsung; and from China, the University of Shanghai.

Citation frequency of scientific articles

In addition to the sheer number of journal articles in the Web of Science databases, in the course of our investigations we also analyzed the citations of these articles. An analysis of the frequency of citation allows us to not only look at quantity but also aspects of importance and quality of the publications involved. The frequency with which an article is cited can serve as a measure of its quality or scientific importance. A query of the Web of Science Citation Report listed a total of 72,823 citations for 4,878 publications that were found for biomimetics in the narrow sense, which corresponds to an average of 15 citations. Of these, 46% of the citations occurred for publications from the USA; Japan, Germany, and Great Britain each had a 10% share. With France taking 6.7% and China only 2.5%, these countries covered 86% of all citations. American (21.2) and British (20.8) publications account for being the most frequently cited followed by Germany with 16.6 citations per publication, thus evidencing a greater scientific "impact." While France and Japan came in just below the worldwide average, Chinese publications clearly showed the lowest value of the countries considered.

Scientific impact

If we take the expanded search terms *(learning from nature)*, the picture changes only slightly. However, it is noticeable that publications from Germany are, on average, cited less often than either French or English publications, even though German publications make up the second largest share of all publications (13%, well ahead of 7% from Great Britain and 7.5% from France).

92 These statements are supported by analyses that for the sake of brevity are not depicted here.

Figure 22 — American research networks for "biomimetics in the narrow sense"

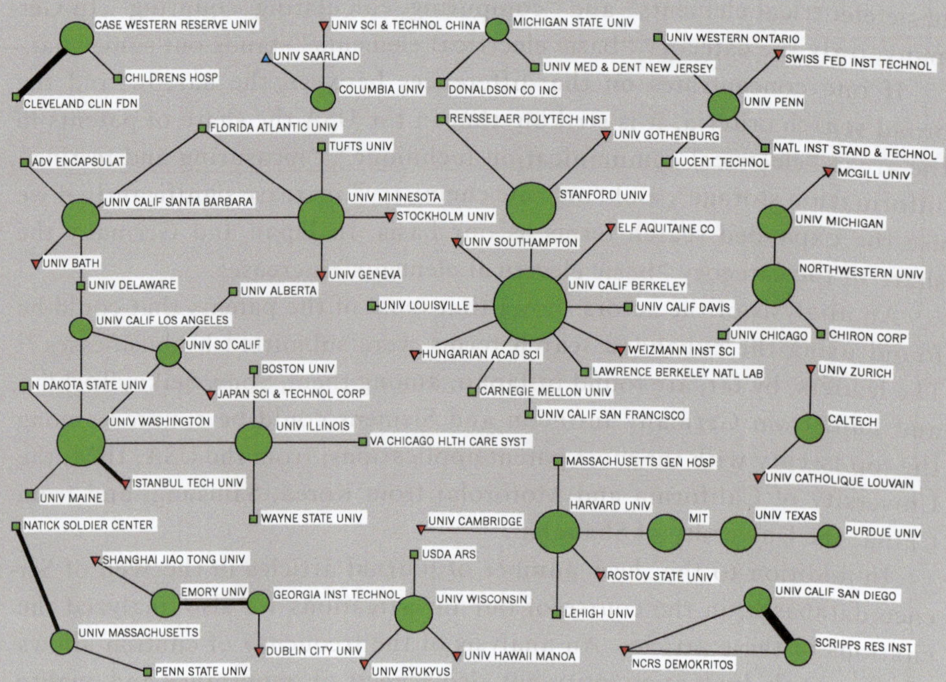

[Source: Compiled using data from the Web of Science databases.]

Figure 23 — American research networks for "learning from nature"

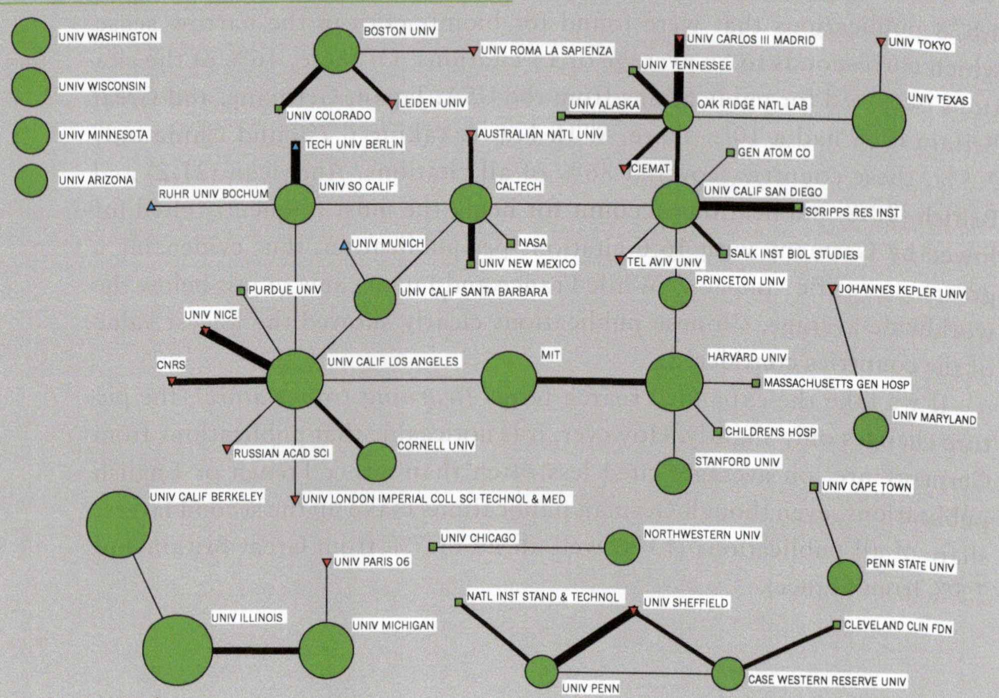

[Source: Compiled using data from the Web of Science databases.]

Network analysis

Identification of cooperations on joint publications

The network analysis is based, once more, on an analysis of publications for *biomimetics in the narrow sense* and *learning from nature* in the literature databases of the Web of Science.[93] For each country, the twenty-five leading institutions with the most publications were more closely analyzed and exclusively those cooperations depicted that are documented by at least two joint publications (for *biomimetics in the narrow sense*) or at least five joint publications (for *learning from nature*). The results of the analysis were then graphically presented using the software program NetDraw; thus, the size of the hubs reflects the number of publications and the thickness of the connecting lines, the number of joint publications.[94] Thereafter we depicted the research networks for *biomimetics in the narrow sense* and *learning from nature* for the USA, China, France, Japan, Great Britain, and Germany.

4

Cooperations in the USA

In the American research landscape for *biomimetics in the narrow sense* (Figure 21), we find that each of the 25 leading institutions cooperates with at least one other institution. The strongest collaboration occurs between the Scripps Research Institute and UC San Diego, with a total of thirteen joint publications. Remarkable is furthermore that the institution with most publications, the University of California, Berkeley, cooperates with a total of ten other institutions, whereby the number of joint publications does not exceed three.

In the research landscape for *learning from nature*, (Figure 23) the networking of the actors is also striking. Only four universities publish without collaborating with other research institutions. Twelve of the institutions with the greatest number of publications are linked together in a research network encompassing the universities of Harvard, California (San Diego), Caltech, California (Los Angeles), and MIT.

As a whole it can be seen that the landscape of biomimetics actors in the USA is particularly distinguished by relatively large number of institutions that extensively publish and are networked.

Cooperations in China

An institutional analysis of the results for China (Figure 24 and Figure 25) shows that the research for *biomimetics in the narrow sense* as well as for *learning from nature* is concentrated in the Chinese Academy of Sciences

93 For the analysis of *biomimetics in the narrow sense* we searched the time period 1990–2005 using the following search terms: *bionic* or *biomimetic*; for the analysis of publications in the sense of *learning from nature* the same time period and the following search terms were used: *bionic**, *biomimeti**, *bio-inspir**, *self-repair**, *self-heal**, *self-organi**.

94 Node symbols: circle = institution is one of the 25 leading institutions in the respective country; square/blue = domestic cooperation partner, not among the top 25; triangle/red = foreign cooperation partner (if any); green = cooperation partner from Germany. Strength of relationship: Number of joint publications for the analysis *bionics / biomimetics in the narrow sense* (values for the analysis *learning from nature* in parentheses) 1pt = 2–3 (5–6), 2pt = 4–6 (7–8), 3pt = 7–10 (9–10), 4pt = more than 10 joint publications.

Figure 24 — Chinese research networks for "biomimetics in the narrow sense"

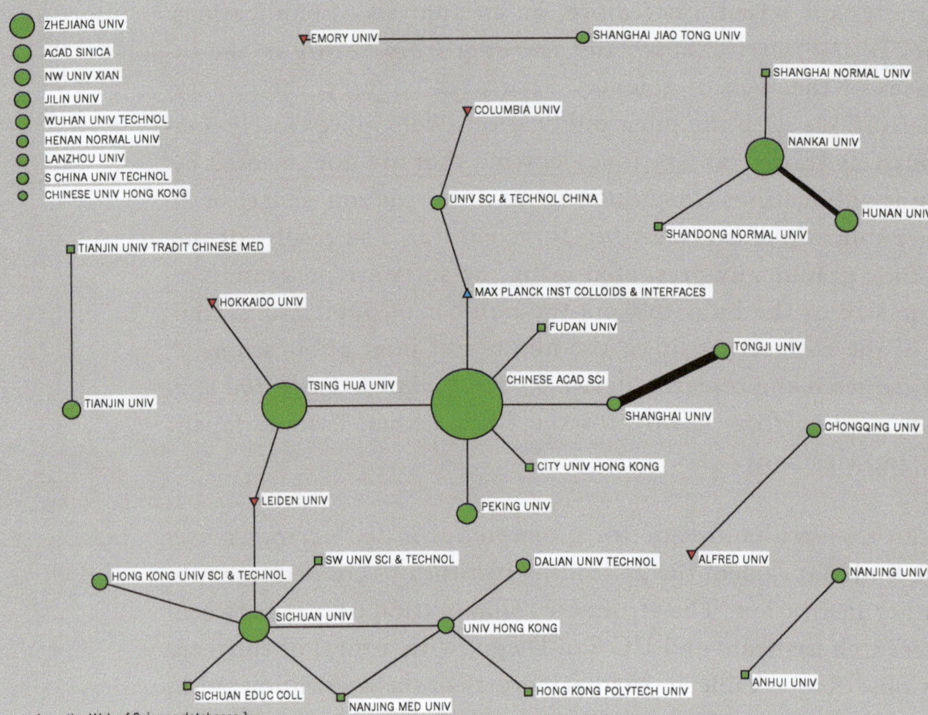

[Source: Compiled using data from the Web of Science databases.]

Figure 25 — Chinese research networks for "learning from nature"

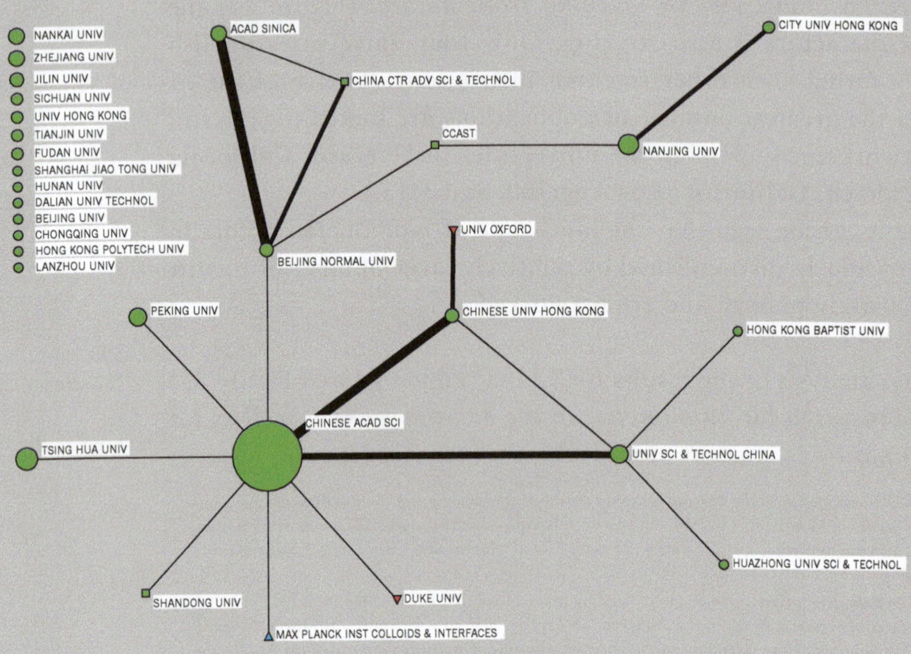

[Source: Compiled using data from the Web of Science databases.]

(CAS, Zhongguo Kexueyuan). The CAS is the most important science and technology advisory body to the Chinese government. In addition to these advisory tasks, it manages more than 100 research institutes.[95] To an extent unlike in any other country, the research activities for both *biomimetics in the narrow sense* as well as *learning from nature* are concentrated here. The CAS, however, is also a publication participant in almost 20% of all entries of Chinese institutions in the Web of Science. The CAS publications for biomimetics and related terms make nearly 17% of all biomimetics publications from China.

The network is centered on the Chinese Academy of Science, the institution with by far most publications from the country (332 publications, approximately one fourth of the countrywide publications). The Chinese Academy of Science cooperates with a total of eight partners (two of them from abroad). The strongest cooperation exists between the Academy Sinica and Beijing Normal University (twelve joint publications).

More than half of the leading 25 institutions publish without the cooperation of other institutions.

Cooperations in France

In France (Figure 26 and Figure 27), similar to China, one actor stands out in particular, the Centre National de la Recherche Scientifique (CNRS), which produced most publications for *biomimetics in the narrow sense* (60 entries) as well as for *learning from nature* (287). The research landscape for biomimetics in the narrow sense is grouped around the three central actors, CNRS, University Paris 05, and University Paris 06, which collaborate closely with the College of France (four joint publications).

In the diagram for *learning from nature* (Figure 27) the strong networking also stands out. State institutions such as CNRS, the Commissariat à l'Énergie Atomique (CEA), the Center for Nanoscience (CENS), and numerous universities stretch out to form by virtue of close cooperation through joint publications a central network with 19 participating institutions. There exist three further networks, each with three participating institutions.

Cooperations in Japan

In Japan, six of the twenty-five leading institutions have two or less joint publications with other institutions in the area of *biomimetics in the narrow sense* (Figure 28). The strongest visible cooperations exist between the Nara Institute of Science and Technology and the Toyobo Research Center Co. Ltd. and also the Kyushu Institute of Technology (nine and eight joint publications, respectively). The leading biomimetics research institution, Kyoto University, works together with eight further institutions, in particular, with the Kyoto Institute of Technology and the Matsue National College of Technology (each six joint publications).

95 *Source:* VDI-Technologiezentrum 2006, 18.

Figure 26 — French research networks for "biomimetics in the narrow sense"

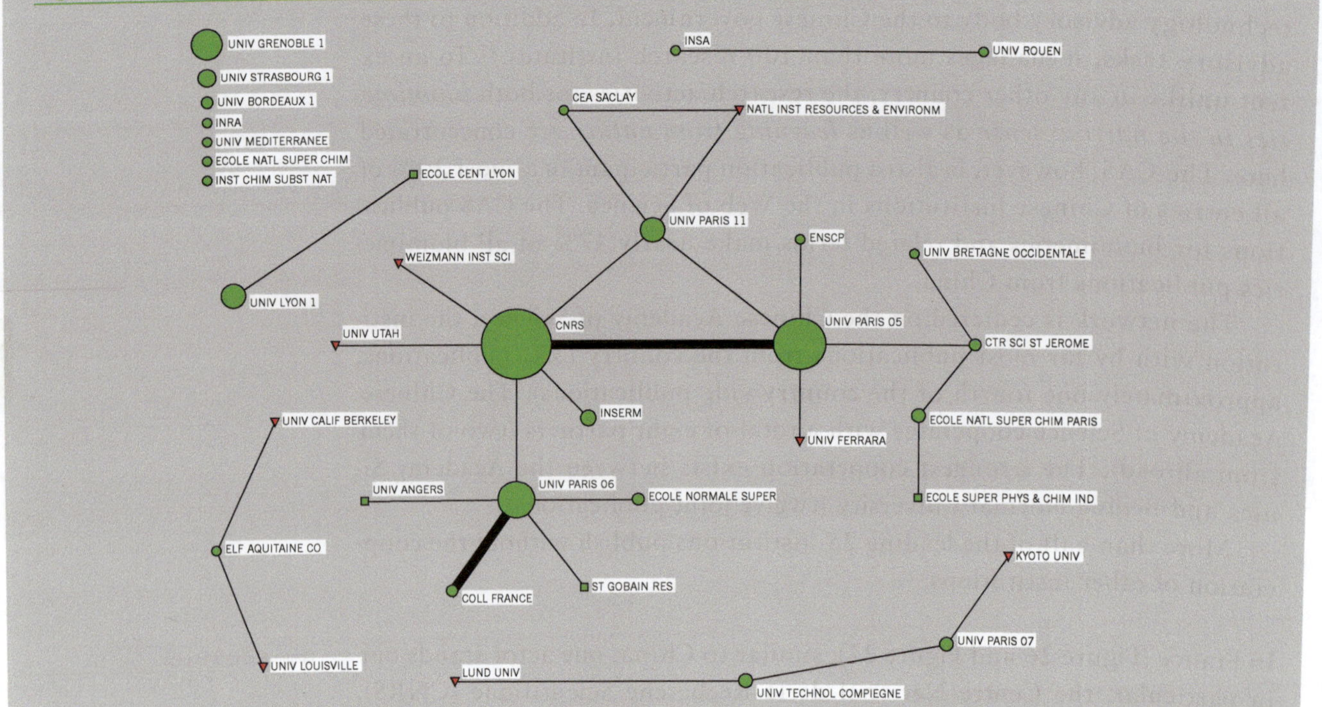

[Source: Compiled using data from the Web of Science databases.]

Figure 27 — French research networks for "learning from nature"

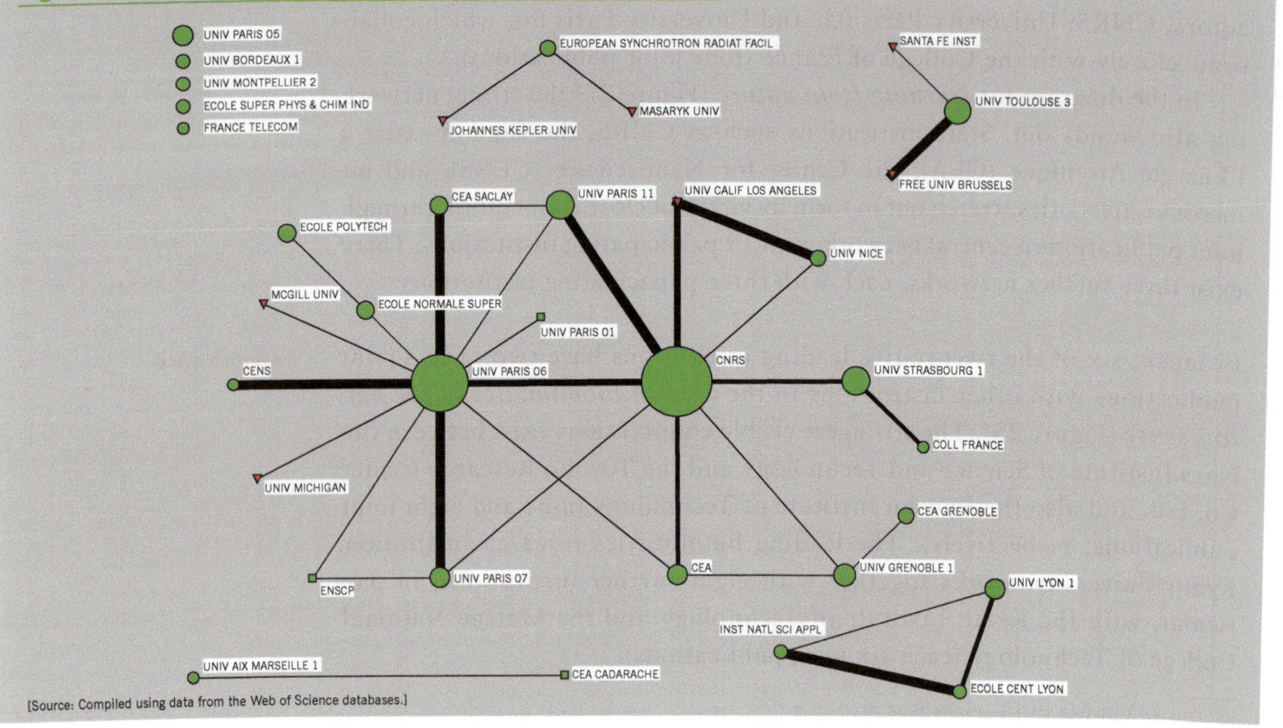

[Source: Compiled using data from the Web of Science databases.]

In the area of "learning from nature" (Figure 29), the University of Tokyo, with a total of eight cooperation partners and an average of more than six joint publications with each of these institutions, enjoys a key position among the Japanese networks. In addition, the network of the universities of Kyoto and Osaka and the Kyoto Institute of Technology play a significant role within the Japanese biomimetics research landscape. There exist five further smaller networks with various features of diverse characteristics in which participate the Kyushu Institute of Technology as well as the Nagoya University.

Cooperations in the UK

In the British biomimetics research landscape for „biomimetics in the narrow sense" (Figure 30), three centers can be determined from which three cooperative relationships originate: the universities of Southampton, Cambridge, and Oxford. The University of Southampton has published jointly with the University of Nottingham (4); the University of Oxford, with the University of Loughborough (6) as well as the University of Dundee (4). Of note is that the universities of Bath and Reading, which hold a central role in the British biomimetics network,[96] are certainly recognizable here as actors, but do not participate in the three most active (in terms of publications) networks. This, too can serve as evidence that those institutions that explicitly acknowledge biomimetics or bionic (and participate in such networks) do not show up as particularly strong in publications in international journals, instead publication and patent analyses tend to pick up publications and institutions that are working in the broader periphery of biomimetic approaches.

Cooperations in Germany

See figures 9, 11, chapter 4.5

A look at the research landscape with respect to *learning from nature* reveals that in Great Britain the largest share of the leading institutions (18) publish separately; this includes the universities of Bath and Reading, sponsors of the BIONIS network. One single network of more than three actors, centered on the University of Oxford, can be identified. British institutions forge the strongest cooperative arrangements with foreign institutions.

The most striking aspect of the German research landscape with respect to both biomimetics variations (Figure 9 and Figure 11) is the large number of institutions among the most active publishers that only publish separately (13 for *biomimetics in the narrow sense* and 12 for *learning from nature*). The field of "biomimetics in the narrow sense" consists of several small networks (Figure 9). The most fruitful collaborations are those between the Max Planck Institute for Polymer Research and Uni-

96 *Source:* Web page of the Biomimetics Network for Industrial Sustainability on the server of the University of Reading, http://www.extra.rdg.ac.uk/eng/BIONIS/what_is_bionis.htm.

Figure 28 — Japanese research networks for "biomimetics in the narrow sense"

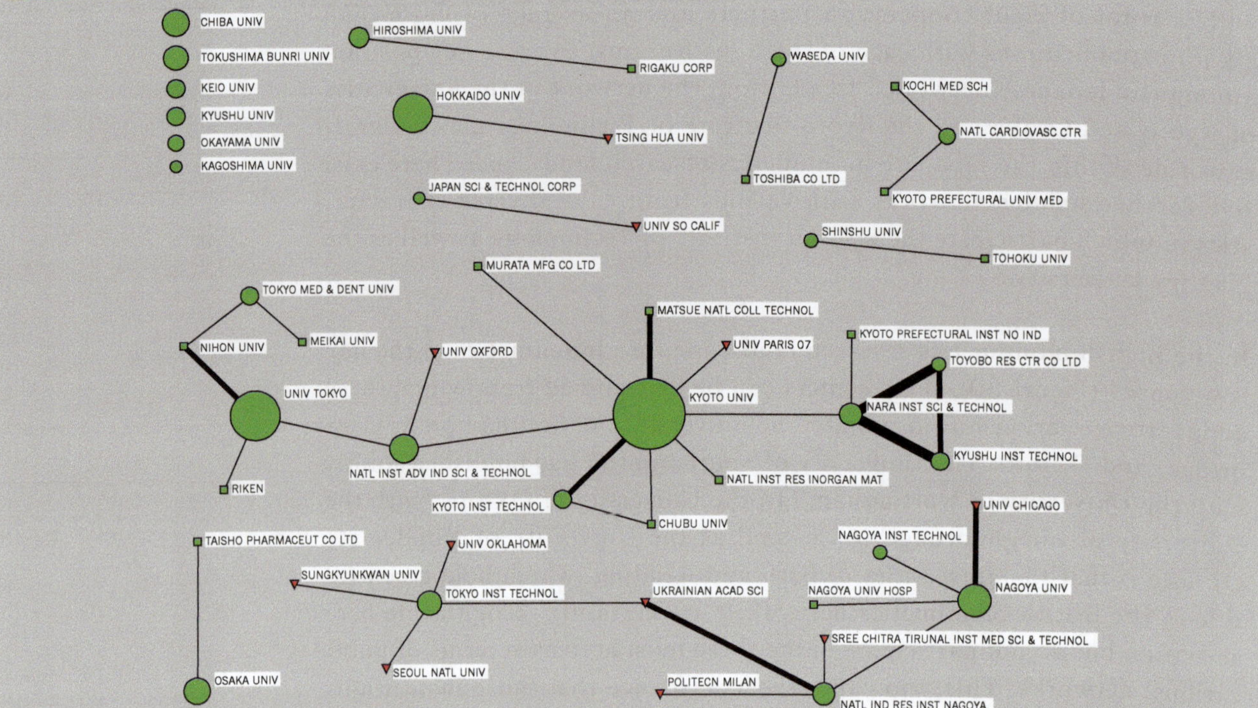

[Source: Compiled using data from the Web of Science databases.]

Figure 29 — Japanese research networks for "learning from nature"

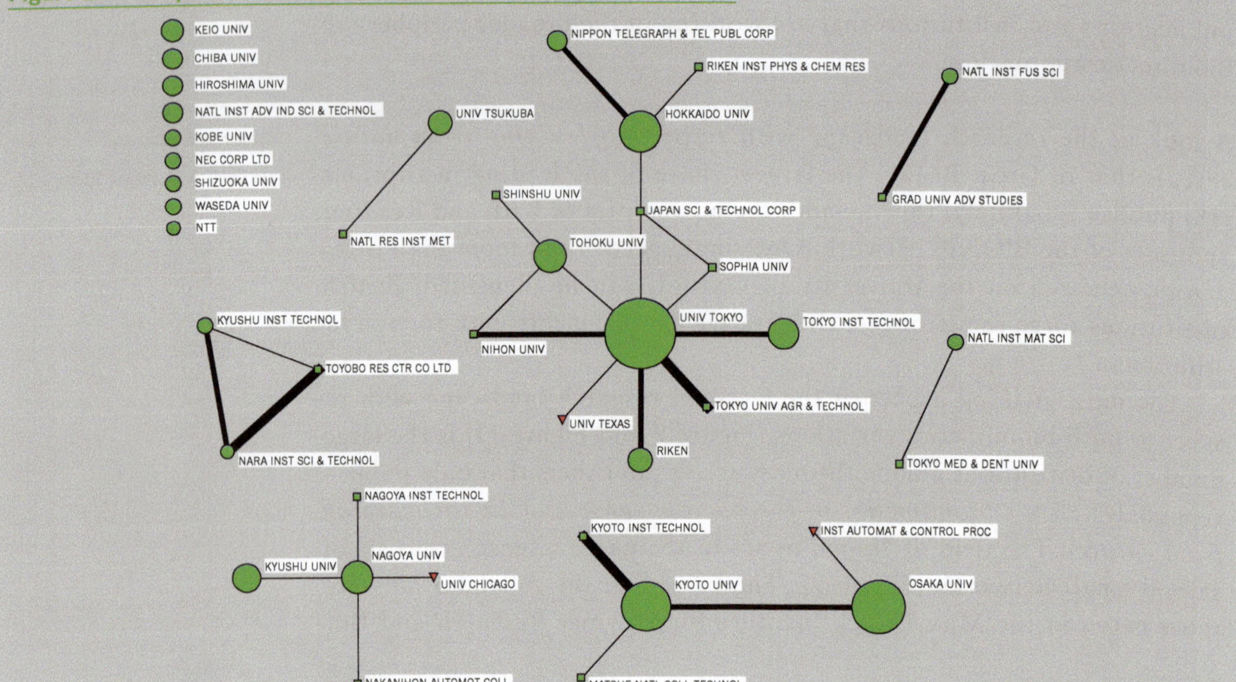

[Source: Compiled using data from the Web of Science databases.]

Figure 30 — British research networks for "biomimetics in the narrow sense"

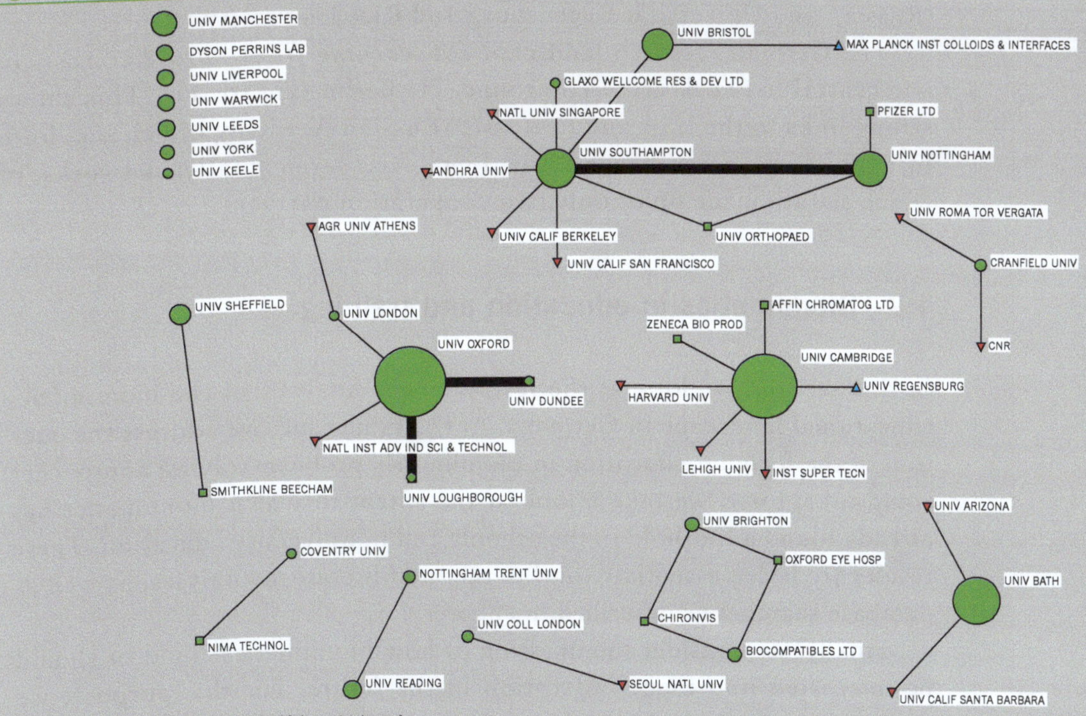

[Source: Compiled using data from the Web of Science databases.]

Figure 31 British research networks for "learning from nature"

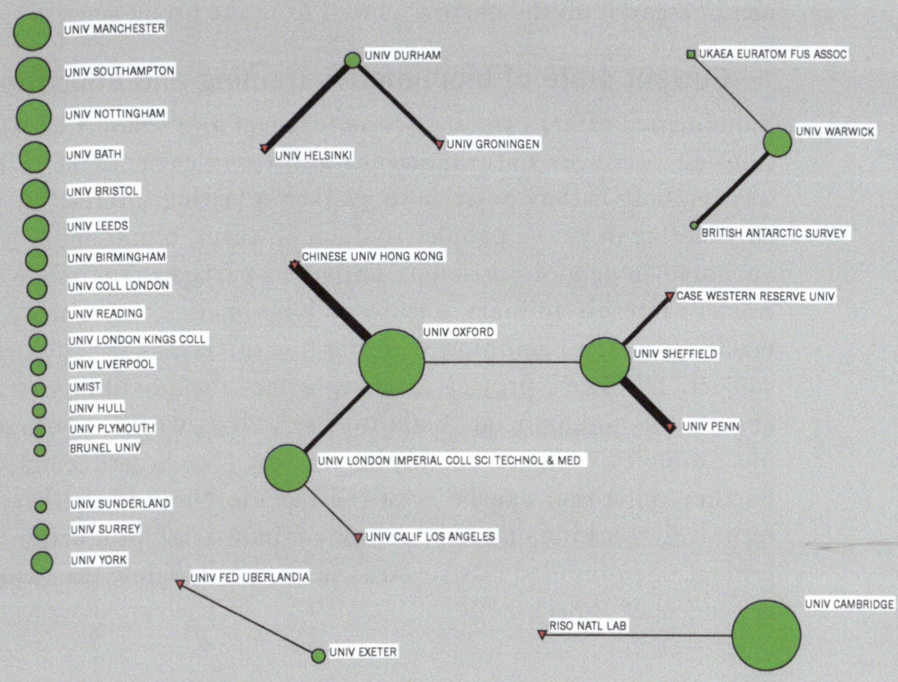

[Source: Compiled using data from the Web of Science databases.]

versität Mainz, Universität Ulm and Lomonosov Moscow State University (Russia), and Universität Regensburg and Rice University.

The German research landscape for *learning from nature* consists of one central network centered around TU Berlin (Figure 11). This shows strong links to the USA and to Russia (Russian Academy of Science, Ioffe Institute). Above and beyond this, there exist seven smaller networks, of which six are made up of only two cooperation partners.

4.7 Biomimetics in education and training

In this section, we look at the results of the analysis of the biomimetics educational landscape in Germany. At the same time, we address the question of the form an education in biomimetics problem-solving approaches could take. First, we take a look at the extent to which biomimetics has already found a foothold in the schools and in university education. These results are based essentially on Internet and literature analyses, as well as database searches as described in chapter 4.5.

See chapter 4.5

Finally, we consider the question of how biomimetics could or should be integrated into higher education in the future. For this purpose, we drew on the results of literature analyses dealing with training in the complexity sciences or "enabling technologies," for example, the nanosciences. This section is supported by expert interviews with scientists doing biomimetic research (in the narrow as well as in the broad sense).

Current state of biomimetics training and education

Biomimetics as intriguing approach to sciences and technology in schools

Biomimetics offers primary schoolteachers and schools excellent opportunities to convey natural science and technical content in an exciting way, making it thus possible to awaken a lasting interest for nature and engineering in young people. To date, however, biomimetics is not firmly anchored in school curricula. This is in part perhaps related to the extremely interdisciplinary nature of biomimetics, in contrast to school teaching, which beginning with the fourth class is strictly organized by subject. However, project teaching offers the possibility of practicing abstraction and associative abilities as well as working with analogy using examples designed to inspire interest (Nachtigall 2002, 447–448). Further skills that can be acquired include "interdisciplinary scientific technical thinking, understanding of industrial production processes, critical open-mindedness towards new technologies, teamwork, and independence" (Speck 2006).

The members of the biomimetics community, therefore began deliberate to introducing biomimetics into schools already some time ago.[97] In 2006 alone, a biomimetics textbook for senior-level instruction (Hill 2006) and a children's book on biomimetics in the "What is that?" series (Zeuch 2006) were published. C. Mattheck, in particular, attempts to make the content of his biomimetics research and applications easily accessible (Mattheck 1999, 2003, 2006). Alongside Hill (didactics of technology, Uni Münster) the Plant Biomechanics Group Freiburg is developing simple and inexpensive breadboard construction kits in the area of biomimetics/bionics with the goal of deploying these in schools or in adult education (Speck 2006).

Biomimetics exhibitions, especially for children and teenagers

Exhibitions play an important role in the public mediation of biomimetic knowledge, particularly for children and young people. For example, under the title "Bionics – Future Technology Learns from Nature," two traveling and one permanent exhibition were developed by the Landesmuseum für Technik und Arbeit in Mannheim (State Museum of Technology and Trades in Mannheim) in cooperation with Siemens Forum.[98] The Natural History Museum Solothurn prepared the exhibition "High-Tech Nature – Learning from Nature," which was shown in the Natural History Museum of the Palatinate, among other locations. In 2006, the Deutsche Bundesstiftung Umwelt (DBU, Osnabrück), in cooperation with BioKoN e.V., opened the exhibition "Inspiration Nature – the Bionics Patent Workshop," which is specifically aimed at young people in order to awaken their interest in the science of biomimetics (according to the organizers).[99] The exhibition was conceived as a travelling exhibition that would be lent out to schools beginning in January 2007. In June 2007, a series of science days on the topic of biomimetics took place in Luisenpark Mannheim.[100]

Comparison to initatives related to nanotechnology

However, a comparison with initiatives that attempt to bring school children closer to nanotechnologies, for example, demonstrates that there are still further approaches worth following. For example, via an Internet Web site (www.lehrer-online.de), study units on nanotechnology have been prepared and made available online (Humpl et al. 2006), and the association NanoBioNet developed the NanoToolBox, a collection of materials for experimentation, in order to make nanotechnology more accessible.[101] VDI, the Association of German Engineers, has published a study guide, and the Federal Ministry of Education and Research (BMBF) started the in-

97 See http://muenster-uni.biokon.net/; available on this home page are a collection of example experiments for classroom teaching, a series of work materials for download or available by order, information on the integration of biomimetics in teaching.

98 See http://www.landesmuseum-mannheim.de/tp10301.htm; http://www.landesmuseum-mannheim.de/bionik/start.htm

99 http://www.dbu.de/123artikel26029_335.html?PHPSESSID=e5cebf9f76a93ac365e87ea7e5b43c09

100 http://www.explore-science.info/

101 See http://www.lehrer-online.de/dyn/9.asp?url=511473.htm

Table 13 — Universities offering courses in biomimetics

No.	University	Course offerings
1	RWTH Aachen	lectures and seminars
2	Technische Universität Berlin	lectures and seminars, compulsory optional subject
3	Universität Bielefeld	lectures and seminars
4	Universität Bochum	lectures and seminars
5	Universität Bonn	lectures and seminars
6	Hochschule Bremen	International course of study in bionics (ISB)
7	Universität Bremen	lectures and seminars
8	Technische Universität Darmstadt	Interdisciplinary area of specialisation in biotechnology
9	Technische Universität Dresden	lectures and seminars
10	Universität Freiburg	university research project
11	Technische Universität Ilmenau	area of specialisation in biomechatronics
12	Hochschule Magdeburg-Stendal	lectures and seminars
13	Universität Münster	lectures and seminars
14	Hochschule Saarbrücken	lectures and seminars
15	Universität des Saarlandes	courses and lectures (biomimetics course of study – technical biology discontinued in 2002)
16	Universität Tübingen	lectures and seminars

[Source: authors' own research]

itiative Nanotruck (http://www.nanotruck.de) in 2004 in order to make the technology available to a broad audience, particularly to school children.

Biomimetics in university education

In the area of higher education, one finds a mixed bag with respect to biomimetics in teaching. Lectures, seminars, electives, core courses or degrees could only be identified at sixteen universities in Germany (see Table 13). The overwhelming majority of universities offers only single, mostly optional courses. At only five universities does biomimetics play an important role and only one university offers a full-fledged degree in bionics (HS Bremen). The bionics study program set up by Prof. Nachtigall in Saarbrücken was abandoned following his retirement.

Identification of biomimetics related lectures

However, a word of caution about a possible over-interpretation of the conclusions to be found in Table 13 is to be said. The difficulty with enquiries at universities about possible biomimetics lectures and seminars is that – as with all efforts to identify biomimetics research and work – a number of instructors offer courses with biomimetics content that is not identified as such; the exact number of such cases is extremely difficult to judge. Thus, one must assume that in the table above, all lectures and seminars belonging to *biomimetics in the narrow sense* are represented, but most likely only a very small number of those fall under *learning from nature*. For example, at TFH Wildau, the subject Biosystems Technology/ Bioinformatics is taught; included in this are courses on biosensors.[102] The Department of Neurobiology at Universität Ulm offers lectures on sensor technology, motion control, and functional morphology; also attached to the department is Fritz-Olaf Lehmann's BioFutures Research Group, within which research into the flight characteristics of insects is done. However, none of the courses there are categorized explicitly as either biomimetics or bionics. Therefore, this table can only provide a suggestion concerning current teaching.

What form should biomimetics in education and training take?

The analysis of the central developments and technological perspectives in biomimetics has already shown that it is reasonable to distinguish between two ranges or interpretations of biomimetics: *biomimetics in the narrow sense* and *learning from nature* (see in particular the exposition at the beginning of chapter 4 as well as chapters 2 and 3). In our analyses of the educational landscape with respect to biomimetics, we concentrated on the narrow interpretation of biomimetics. A broad interpretation of biomimetics could also record those approaches, in which an attempt is made to learn from nature without this being explicitly mentioned. However, as long as no databases exist in which queries using the respective search terms can be made, such approach would require efforts above and beyond the scope of this study.

See chapters 2,3 and 4

As already mentioned, biomimetics in the broad sense overlaps in numerous areas with the trend towards converging technologies. It therefore makes sense to look at examples of biomimetics in the course of other investigations of enabling technologies and general purpose technologies particularly to develop a corresponding broader educational concept. The converging technologies will bring about enormous social and economic changes (see BMBF 2006b). To a certain extent, the development of biomimetics itself has anticipated this trend. Biomimetics, therefore, can be understood as a sort of testing ground upon which suitable approaches and concepts for dealing with converging technologies can be developed. Thus, above and beyond the narrower issue of a biomimetics curriculum, the question that needs to be addressed is: How can we best prepare the next generation of scientists to developing and dealing with enabling and general purpose technologies?

»Learning from nature« and »converging technologies« in education

Minai et al. (2006) make the case for a new paradigm within the engineering sciences, which should then be incorporated into the higher education curriculum. The central challenge that engineers will face in

New paradigm in dealing with complexity

102 See http://www.tfh-wildau.de/bio/

the future will be coming to terms with the growing complexity of technology. The authors distinguish here between "complex" and "complicated" systems (ibid, 2). The latter are made up of single elements assembled together, where it is the number of elements that determines the degree of their complexity or intricacy. Complex systems, on the other hand, represent self-organising, emergent properties, which is why such systems cannot be reduced to the number of their individual elements. This distinction is important because according to Minai et al. (2006), engineers (and also their education and training) were previously geared to the development of ever more complicated technical systems. The goal of the classic engineering science approach consists of creating technological applications that are directly controllable by the developer, whose properties can be explicitly described, and whose modes of action thus are predictable (ibid, 8).

The authors see the future of the engineering sciences in the exploration of more complex systems with its goal being the development of technical applications capable of dealing with complex properties. To do this, it will be necessary for engineers to become involved with self-organizing, evolving, self-repairing and adaptable properties. The authors call for "a drastic re-evaluation of many fundamental assumptions and methods of the classical engineering paradigm" (Minai et al. 2006, 4). As an initial approach, in order to realize such a change in the assumptions and methods of the classic paradigm, the authors suggest, would at the least consist in augmenting the contents of the higher education engineering science curriculum to include courses on complex systems.

Enabling technologies and general purpose technologies

Deliberations on the training of skilled personnel capable of working with enabling technologies and doing research are also found in studies that deal with questions of training and education in the area of nanotechnology, which is also numbered among the enabling and general purpose technologies (e.g. Humpl et al. 2006). The study "Trendqualifikationen Nano" ("Trends in Nano-Qualifications and Skills", Abicht et al. 2005) attempted to determine which professional qualifications and requirements are necessary in the area of nanotechnology; this was done by means of surveying of businesses, research networks and expertise centers, as well as research and educational institutions. The results show that in most of the skill trends,[103] a mixture of knowledge from physics, chemistry, biochemistry, biology (and other subjects, depending on specialisation) is required. This result likewise points to the converging technologies and the necessary analogies of the various separate science disciplines.

Biomimetics and nanotechnology in education

Humpl et al. (2006), in their study "Aus- und Weiterbildung in Nanowissenschaften und Nanotechnologien – Bestands- und Bedarfserhebung

103 Nanochemistry, nano-analytics, nanobiotechnology, nano-optics, nano-electronics

in Österreich" ("Education and Continuing Education in the Nanosciences and Nanotechnologies – Survey of the Present State and Future Needs in Austria") also investigated measures in Germany designed to promote education and training, as well as continuing education in the nanosciences and nanotechnology. Of interest here is that one part of the already mentioned NanoToolBox consists of an experiment illustrating the so-called lotus effect.[104] The lotus effect thus serves as a paradigmatic example for the successes within biomimetics and the nanotechnologies. Likewise, the brochure published by the BMBF (2006a), "Nanotechnology – Innovations for the World of Tomorrow," makes multiple use of biomimetics topics and illustrations (e.g. the gecko foot – use of Van der Waals forces). This, too, emphasizes the proximity of the two research approaches and shows that the efforts being made to establish the nanosciences, nanotechnologies, and biomimetics in the broad sense can be linked together.

4

Looking at the establishment of the nanotechnologies in the higher education curriculum, the authors of this study make a plea for the introduction of independent courses of study in the area of the nanosciences. Furthermore, they call for increased support of diploma theses and dissertations in this area as well as increased informational campaigns. The following quotation from one of their interviews is typical:

> *"More and more physicists, thanks to the nano-trend, want to go in the direction of bio, although they know nothing about it. I mean, only those who truly understand interdisciplinarity should follow this trend. Many of our people lack the connecting knowledge between biology and the engineering sciences. A combined course of study would be fantastic. I am invited all over the world because I can explain biology in the language of the engineers. If one wants to do nanobio, one must learn a great deal in bio and nano, and not many physicists are doing that."*
>
> (HUMPL ET AL. 2006, 174)

Necessity of changing curriculum contents

On the whole, it can be seen that all of the studies dealing with the establishment of the enabling technologies in the educational system point out the necessity of a change of content to the pursuit of self-organising, evolving systems; the necessity of a stronger linking of the individual disciplines such as physics, biology, chemistry, IT; and the necessity of introducing new courses of study.[105]

Interdisciplinarity

Although we can follow the first part of these statements without any further ado, we must be somewhat more careful with recommendations for

104 Humpl et al. 2006; see also: http://www.lehrer-online.de/dyn/9.asp?url=511473.htm

105 Such recommendations are also found in: Bruck et al. 2004, Bretthauer 2005, Rühle et al. 2001

the establishment of independent courses of study. Interdisciplinarity must be built upon disciplinarity. There are after all many good reasons why the sciences have organised themselves into and developed as disciplines. This includes the development of disciplinary depth and specialisation, which play a role just as important as the peer review function. Biomimetics is an extremely broad "field," which, along with biology, builds on almost all scientific-technical fields of application. Biomimetics is probably even broader in its make-up than the just discussed nanotechnologies and more comparable to the converging technologies. Viewed in this manner, biomimetics is really only teachable and learnable in interdisciplinary (work) relationships.

Basic vs. post-graduate studies

The question remains as to how biomimetics (outside those few rare ideal constellations) can nonetheless become established in teaching in such a way that appropriate foundations can be laid down for the skills needed to achieve more effective and more efficient transfer of biological phenomena into technical systems. One possibility could be to create a new independent (basic) degree in which the relevant content necessary for biomimetics in the broad sense would be imparted. That would mean that in this degree greater emphasis would be placed on teaching content such as self-organisation. Any degrees would also have to be organised in an interdisciplinary fashion, as they would have to convey knowledge from the fields of chemistry, physics, biology, and IT (and possibly further disciplines). What we view as the more likely possibility would be postgraduate studies (master's) and graduate colloquia on biomimetics approaches based on a solid natural or engineering sciences fundamental education. The current transition to bachelor's and master's degree programs taking place not only in Germany offers good conditions in this respect for introducing the commensurate biomimetics postgraduate studies programs.

Studying biomimetics at Bremen University of Applied Sciences

For now, it is not yet possible to determine which model of biomimetics in education and teaching is the most effective. The full-term biomimetic course of study at Bremen University of Applied Sciences (Hochschule Bremen) represents a success model that should be looked at and scientifically evaluated. Up to now, most of the experts interviewed, however, have been more in favor of the introduction of postgraduate studies. They argue that the understanding of biomimetics approaches to finding solutions for the technological challenges presumes to a considerable extent a basic foundation in the sciences. Young researchers require, in particular, a solid education in one of the natural sciences, such as biology, chemistry, or physics. Building on this qualification, it then makes sense to turn to biomimetics research and become involved in topics from other disciplines. Some of the

Duration of studies

scientists that were interviewed were of the opinion that a longer period of training and education is necessary in order to train scientists and researchers to be able to work on biomimetics projects in the broad sense, because of the need to acquire knowledge from several disciplines. Attempts to impart this knowledge in a shorter time would only lead to a lowering of the educational standard and a lowering of the chances to work at the forefront of biomimetics research.

Experts surveyed in the course of the Swiss study "Future Engineering" were in agreement that "engineers with a strong foundation in the discipline" are essential ("Zukunft Engineering" – Kiener 2005, 12). This point of view is also confirmed by the VDI/VDE Society for Measurement and Automatic Control task force, which grappled with the problem of the future of its branch (Bretthauer 2005). They too, emphasise that automatic control engineering will merge together with other disciplines (computer science, biology) and that this must also be taken into account in the training; in fact, "at the center is a broad, solid fundamental education in mechanical, electrical, and process engineering" (ibid., 157).

Biomimetics includes many individual disciplines

Inasmuch as biomimetic research draws in many different individual disciplines that are not usually closely linked to one another – for example, biology and information technologies, or, in the case of self-organisation, the humanities and social sciences (see, for example, Fromm 2005), the idea of funding project teams on biomimetics topics seems promising. These teams should consist of scientists and researchers from various disciplines working together on a joint biomimetics research project. Younger researchers specializing in biomimetics in their doctoral work could also be integrated into such projects.

This proposal can also be found in the Max Planck Society white paper on the future of the material sciences (Rühle et al. 2001). In the report, the authors emphasize the importance of coooperation between physicists, material science researchers, chemists, and biologists (ibid., 306) for the development of intelligent, biomimetic, self-ordering materials. The authors see possible pathways to more interdisciplinary research in the creation of research institutes or laboratories in which researchers of various disciplines work, as well as in the networking of researchers from different fields (ibid., 307).

4.8 Tentative conclusions about topics, actors, and research networks

Biomimetics community is relatively small

Although Germany is seen as a centre of biomimetic research and development, the number of actors in the German biomimetics landscape with an explicit biomimetics self-image is relatively small and also very heterogeneous in its make-up. A total of approx. 170 actors could be identified in the course of our investigation that in some form or another show a connection to biomimetics. However, the individual actors differ greatly in to the manner and intensity in which they do biomimetics work. For example, among the universities we find a relatively small number in which biomimetics is institutionalized as a branch, field of study, or cross-disciplinary research network. In the majority of the cases, it is only individuals at the respective universities that are involved in biomimetics. In the non-university research institutions, the situation is similar. The core of the biomimetics actors, however, can be divided into two groups, according to whether they are organized in one of the two German biomimetics networks, BioKoN or Kompetenznetz Biomimetik. Only a portion of the biomimetics actors are organized in the two research networks. Commercial firms, in particular, have been underrepresented in the networks to date. The publications and patent analyses do not allow the identification of any particular main areas of emphasis in German biomimetics. This suggests that those actors in Germany who themselves make reference to the terms *biomimetics* and *bionics* are certainly well networked, but convert too little of their networking into international publishing activities or patents.

Intensity of biomimetics activities varies according to actors

Institutionalisation hardly developed

All in all, the circle of those explicitly doing biomimetics or bionics is much smaller than the group of those implicitly doing biomimetics, as the database analysis demonstrated. In a large number of research and development areas, approaches based on biological models apparently have risen independently of one another. With respect to the actors, no technical areas of emphasis can be identified in this broadly interpreted understanding of biomimetics in which biomimetics approaches would be especially pursued. The various actors come out of the most varied disciplines and as a whole cover all those potential areas in which biomimetics might come into play. It would appear, however, that the areas of chemistry and biochemistry, in which (according to the literature research) a vast number of biomimetic approaches are being pursued, tend to be underrepresented to date, at least in the biomimetics research networks dominated by biologists.

Approaches to »learning from nature« are pursued by various actors

Looking at a temporal perspective, the activities of the core biomimetics researchers as well as those focused on a further approach of learning from nature have greatly increased in the past 15 years. The literature database analyses clearly show that not only the use of the core terms *biomimetics* and *bionics* has greatly increased, but also that biomimetic concepts such as *self-healing* or *self-optimisation* are increasingly being discussed in the scientific literature. The increase is not only true in absolute numbers, but also relative to the total scientific publication activities coming out of German research institutions.

Strong increase of biomimetic approaches in past 15 years

At the international level, biomimetics research in the narrow sense is being carried out above all, in the disciplines of medicine, chemistry, bio-

Many hints at connections between biomimetics and convergence technologies

chemistry, and the material sciences. However, one also finds publications from many other areas, such as the computer sciences, polymers, robotics, engineering sciences, etc. The literature and patent analyses based on a broad interpretation of biomimetics *(learning from nature)* provide an empirical confirmation of the hypothesis that in the more widely based biomimetics approaches, the development of converging technologies is being anticipated: the results for publications in which the terms, *biomimetics, bionics, bioinspired, self-organized, self-healing, self-repair* (and their variants) are used, are primarily found in the areas of physics, chemistry, computer sciences, and the material sciences. Although Germany shows a high proportion of publications falling under the broad understanding of biomimetics in the sense of *learning from nature* (more than Japan, Great Britain, and France), publications from Germany are cited less frequently than British or French publications. As for patents, only half as many patents are recorded in Germany as in Japan. Within the countries under consideration, no disciplinary areas of emphasis could be ascertained by means of the literature and patent analyses. It is largely the same disciplines within the respective countries from which the publications in the two biomimetics categories originate.

German biomimetics R&D relatively wide spread...

In the more general approaches of *learning from nature,* one can assume that Germany has a particular biomimetics strength. Germany is very well represented in publications, but has a smaller impact (frequency of citation, particularly in comparison to contributions from Great Britain or France, and a much smaller level of patent activity in comparison to Japan puts this into perspective. Furthermore, looking at a broad interpretation of biomimetics *(learning from nature)*, the German research landscape appears to be rather poorly networked. This can, in addition to the already mentioned lack of integration in the existing biomimetics research networks, also be understood as a ground for the low impact factor (citation frequency), because cooperation partners tend to mutually cite one another.

... though potentials are not yet exhausted

There are signs that the German R&D landscape, in particular when extending to the approach in sense of "learning from nature", cannot yet properly bring its potential strength to play in comparisson to other countries. This can be seen in:

- a lower impact factor (citation frequency) of articles by authors from research institutions in Germany compared to Great Britain or France;
- a weaker networking of the German actors among themselves as compared to other nations that were studied;
- a significantly lower number of patent registrations, at least in comparison to Japan.

Since we can assume that the middle- to long-term potential of biomimetics lies in this broader understanding of biomimetics (and in its third strand of development), there is something to be said for dedicating a large degree of attention to the respective approaches within the information and communications technologies, material sciences, prosthetics and robotics.

Biomimetics as intriguing approach to sciences and technology in schools

In the area of education and training, one can conclude that biomimetics possesses enormous potential to inspire an interest in young people in the natural sciences and technology. This is of special importance now, although the natural sciences are not particularly in demand at school, employees with a good science and engineering education are very much in demand; employees with science and engineering training are a prerequisite if companies in industrialized countries are to compete on the international market. The present lack – and likely even greater lack in the future – of scientists and engineers could be successfully counteracted with the help of biomimetics content in the curriculum – in the primary as well as secondary schools. Members of the (narrow) biomimetics community are already taking advantage of these opportunities and should be supported in this.

Biomimetics at university

The situation in the area of higher education, on the other hand is more difficult to assess: only a small number of universities presently offer explicit biomimetics courses and lectures, which does not necessarily rule out the possibility that many more chairs are dedicating their efforts to biomimetic topics. The challenges to overcome in the further integration of biomimetics in education are interdisciplinarity, and, in particular, the proximity to the converging technologies. A tremendous future potential is generally ascribed to biomimetics (in the broad sense). To deepen its establishment in the universities, experts recommend the introduction of postgraduate courses of study, project support, and the establishment of interdisciplinary research laboratories. In the opinion of experts, the establishment of full-term biomimetics degrees appears to be the less preferable alternative. Nonetheless, before a final assessment, a comprehensive evaluation of the existing course of studies at the Hochschule Bremen should be conducted, where biomimetics (in the narrow sense) is being very successfully taught.

5 BIOMIMETICS AND INNOVATION

»He who wants to see, must have known doubt before.«

ARISTOTLE, 384–322 B.C.

5 BIOMIMETICS AND INNOVATION

5.1 Biomimetics and systems of innovation

Relevance of innovation systems

In the discussions about the capacity for innovation and technological change, the importance of institutional factors was strongly emphasized. The concept of innovation systems, within which these processes of change take place, combines aspects of evolution as well as institutional economics (see, e.g., Edqvist 2001; Freeman 1987; Lundvall 1992; Nelson 1993). The focus on innovation systems and their framework requirements, that is, on the complex interplay of the stakeholders, with numerous recursion loops, was closely related to the rejection of linear phase models of the innovation process. Empirical comparisons have shown that institutional settings and stakeholder networks can differ among different economies and that this can have an impact on the macroeconomic success of innovation. Freeman (1987, 1) defined national innovation systems (NIS) as "the network of institutions in the public and private sectors whose activities and interactions initiate, import, modify and diffuse new technologies."

National

Lundvall emphasizes the importance of "user-producer linkages" in these networks. In order to define them more precisely, he distinguishes between a narrower and a broader interpretation of NIS (Lundvall 1992, 12–13). The narrower interpretation defines the network as being composed solely of those organisations that are actively working on and seeking innovation; this comprises mainly private research and development establishments, universities, and non-governmental technological institutes. The broader definition includes the production system, the market system and the financial system. All these subsystems, he argues, produce and reproduce environments relevant for innovation processes; the same goes for governmental regulation.

Regional

At the regional level, innovation systems are also important (see, e.g., Braczyk et al. 1998); here again one finds the concepts of *cluster* and *innovative milieus* in use. An essential factor behind this is the fact that knowledge spillovers are often regionally limited (often referred to as *localized knowledge*). The necessary trust in many innovation networks is more easily gained through local face to face contacts. All this emphasizes the importance of the stakeholders interplay in the innovation process.

According to Schumpeter, innovation is the successful implementation of new combinations of the existing and the new by the entrepeneur. In other words, new technological applications are generated through the combination of knowledge flows of various knowledge producers (R&D

Research and innovation policy

institutions, producers, users, etc.), which can be described as networks. The form and structure which innovation networks for new technological approaches such as nanotechnology or biomimetics should take in order to prevail against established developmental trajectories and succeed in the marketplace, thus, become an important question for research and development itself as well as research policy and innovation funding. Therefore, policy makers are faced with new tasks, for example, as network initiators, mediators, and sponsors, in order to reduce the time-to-market span or fund new R&D approaches, including R&D approaches to more sustainable development (Kemp 2001, lead market approach – Jacob et al. 2005).

Restricted possibilities of designing innovation processes

However, it should be pointed out that innovation research tends to lower the expectations of far-reaching design optionsby increasingly viewing the development and diffusion of innovations as a complex interactive process, which involves a number of players and is and characterized by multiple feedbacks. Clearly institutional arrangements and stakeholder relationships which influence the development and diffusion of new technologies play a central role in the differences in economic performance of countries, regions and sectors. However, the patterns of innovation processes are subject to dynamic change and innovation systems are constantly being forced to adapt to new challenges such as globalisation, the emerging knowledge society, the distribution of new generic technologies (e.g., information and communication technologies) and the growing public awareness of global problems and (non)-sustainable development.

Incremental and disruptive innovations

Additionally, differentiation with respect to the type of innovation (technological, organisational, institutional, etc.) and particularly its form of progress are appropriate; the latter is accomplished by distinguishing between incremental (continuous) innovations along trajectories and disruptive innovations or path changes. Such disruptive innovations – or those involving a change of path – often cannot be handled by existing innovation systems since they cannot be incorporated structurally (because the knowledge base is not available, for example) or the nature of the new technologies requires considerable organisational changes.

Path dependencies

Innovations, thus, are forced to struggle with enormous systemic inertia and path dependencies. As mentioned before, that alternatives usually must be considerably superior to the system they are intended to replace – and – new innovations must fit into the existing innovation system. Even those "technological superiorities and advantages" highly esteemed by the customer are sometimes insufficient to ensure the actual success of the better alternative, which then loses to those better fitting and already existing innovation systems and trajectories.

It can be assumed that existing innovation systems are normally able to handle regularly recurring, weak impulses, such as changes in demand, "normal" technological progress along the trajectory, or incremental regulatory adjustments, because they correspond to a large extent to the capacity of the existing structure and knowledge base. With respect to more radical innovations, such as those often demanded in connection with biomimetics (as well as with nanotechnology and nanobiotechnology), the question arises as to how innovation systems can cope with more radical, stronger impulses, or even shocks (for example, technological breakthroughs, new markets, crises in existing markets, and new demands or needs from society), whose handling does not call for incremental but rather radical or systemic innovations. Such strong impulses often require more than adaptations to existing systems, that is the creation of new (sub-) systems.

Dealing with radical innovations

Particularities about the German innovationsystem

Although such statements at this general level should be handled with care, it has frequently been stated that the German "national innovation system" is still strongly dominated by the industries and technologies of the second industrial revolution, i.e. the chemical industry, mechanical engineering, and the automotive industry (see Grupp et al. 2002). In addition, the chemical and the industry are characterized, on the one hand, by centralized large-scale enterprises and, on the other hand, by mostly small-scale sub-contractors. Engineering-oriented expertise is furthermore regarded as being typical of the national innovation system in Germany. These core competencies in the German innovation system are based to a great extent on the traditional, technologically oriented economic trajectories. Changes in innovation direction and particularly disruptive innovations can possibly be integrated into existing trajectories – as has happened (mostly) in the case of information technology; they can, however, also lead to a questioning of existing trajectories.

The inertia in the national innovation system has frequently been addressed (e.g., Wieland 2006, Wengenroth 2006); reports on technological performance likewise conclude that Germany is somewhat behind in the field of high technology. Furthermore, it has been noted that the introduction of new technologies in Germany has faced specific obstacles till this day. This affected particularly fields that did not sufficiently fit into the German system, such as biotechnology and genetic engineering, the integration of which has been a problem, particularly for the "chemical paradigm" of the pharmaceutical industry (not only in Germany).

General purpose technologies

The information and communication technologies, as well as biotechnology and genetic engineering are "general purpose technologies" (Bresnahan/Trajtenberg 1996) which can be applied in a variety of fields and are therefore capable of making considerable contributions to processes of change. Presently, comparable innovation effects are expected in particular in the nanotechnology fields (nanobiotechnology). These general purpose technologies are utilized in many fields and industries and are far less centralized than, for example, energy technology (nuclear energy) or aerospace technology. This "decentralism" requires new concepts of funding for research, technology, and innovation that are less focused on public funding of singular lines of technology. Instead, search processes need to be optimized and kept open longer to avoid premature termination processes. The funding of nanotechnology – or more specifically, the funding of innovations based on nanotechnology – offers a current instructional example.

Biomimetics and techno-sciences

Characteristic for biotechnology, genetic engineering, information science, nanotechnology as well as biomimetics is the equiprimordiality or common origin and the particularly close interdependence of technological and scientific development. However, biomimetics differs from these "techno-sciences" especially in its emotional and normative content. In contrast to genetic engineering, biomimetics is not regarded as being harmful to humans or nature. This is probably primarily due to the fact that in biomimetic approaches, the goal is to emulate nature rather than reinvent it (from synthetic chemistry all the way to synthetic biology). Possible arguments for and against these widespread judgments (or prejudices) have already been extensively presented elsewhere (see chapter 2).

See chapter 2

Changes in technology- and innovation policy

Since the end of the 1990s, technology and innovation funding policies have distinctly changed – and not only in Germany. Reasons for this development are, among others, i) the increasing internationalisation of research and business, ii) the increasing importance of research and development (knowledge) for economic development, iii) the structural changes in those fields of technology that are regarded as relevant for innovation and competitiveness, from "large scale technologies" to more multi-purpose technologies (IT, biotechnology, nanotechnology, etc.), iv) the gradual disappearance of the borders between basic and applied knowledge, and last but not least, v) the increasing multidisciplinarity and heterogeneity of knowledge generation and technology development. Innovation policy responded to these changes by funding so-called key technologies and high-tech clusters (in part regional) and by opening up the R&D environment to improve its international competitiveness and attractiveness (see Dolata 2004).

5.2 Biomimetic innovations

So far only limited innovation successes due to biomimetics

In retrospect, the economic-technological successes of biomimetics so far should be regarded with caution. There undoubtedly have been great successes, from airplane wings to "winglets," the hook-and-loop fastener (Velcro), the walking robot and the lotus effect. However, the number of (successful) biomimetic products is rather small (see Table 12 in chapter 4) and in case of some innovation successes there is ongoing debate as to whether the respective solution was actually acquired from nature (for example, Velcro; see Vincent et al. 2006, 473).

See chapter 4, table 12

Table 14 — Examples of biomimetics developments and their phase of innovation (authors' own compilation)

Biomimetic approach	Phase of innovation (year)	Source (not necessarily the innovator!)
Self-cleaning surfaces based on the structure of the surface structure of lotus leaves	Market penetration of various paints and lacquers (since 1999)	Küppers (2004, 320)
Flow resistance–reducing surfaces based on the structure of dolphin skin	Preliminary product available (ca. 2004)	Küppers (2004, 320)
Flow-optimized tubular constructions based on the example of river meanders	Industrial test phase (ca. 2004)	Küppers (2004, 320)
Flow- and weight-optimized automobile: external form based on the example of the boxfish, body components optimisation based on the growth processes of bone and trees	concept phase, research and development (ca. 2005)	Website Daimler/Chrysler (2005)3
Running machine with legs in place of the usual wheels/tracks for forest work	prototype, test phase (1995)	Nachtigall (2002, 190)
Drag-reducing film for aircraft and boats based on the structure of shark skin	Test phase (1990s), thereafter no further development in this area Market introduction for swim suits	Cerman et al. (2005:206ff)
Winglets (specially formed wing-tips on aircraft wings based on the example of bird wings)	Established in the market (since the end of the 1980s)	Cerman et al. (2005, 171–172)
Flapping-foil propulsion for underwater vehicles based on the example of the fish	research and development, initial prototypes (since 1905)	Nachtigall (2002, 49ff.); most recent: Bandyopadhyay, (2005)
Computer-aided method for weight and stability optimisation in the manufacture of mechanical parts based on the growth principle of trees and bone CAO/SKO	established on the market	Nachtigall (2002, 374ff.)
Self-sharpening cutting tools modeled on the incisors of rodents	research and development, prototypes (2006)	Szentpétery (2006, 66)
Multi-story buildings based on the example of plant structures	concept phase (2006)	Gaffron (2006)
Algorithms modeled on the genetic evolutionary principle for the optimisation of tread designs	research and development, cancelled (ca. 2006)	Honsel (2006)
Underwater communication system modeled on the example of dolphin communication	research and development, market introduction (ca. 2006)	Szentpétery (2006, 66); Bannasch/Yakovlev (2006)

Biomimetic approach	Phase of innovation (year)	Source (not necessarily the innovator!)
Self-healing foam coating for light-weight construction frameworks modeled on the example of vines	research and development (ca. 2006)	Szentpétery (2006, 64)
Tactile sensors for robotics applications based on the example of diverse sense organs of various animals	research and development (ca. 2005)	Dargahi/Najarian (2005)
Optical sensors for the navigation of aircraft modeled on the example of the compound eyes of insects	research and development (ca. 2005)	Ruffier/Franceschini (2005)

Many innovations (from the automotive sector, for example) are mainly incremental (innovative improvements), involving optimisations in specific sections of the automobile, from design optimisations to tire tread patterns. The situation is similar for information and communication technology. Many projects receiving lots of public attention, such as the technological implementation of shark skin for flow-optimized and non-stick surfaces or the Berwian wind turbine with rotor blades modeled on the example of bird wings, have not at all prevailed to date. The reasons for the non-adoption of these biomimetic problem-solving approaches have not been systematically examined so far.[106]

Table 14 presents an overview of a selection of attempted biomimetic approaches that have been documented in the literature. We limit ourselves here to biomimetics in the narrow sense, i.e., those situations where the participants make a direct reference to biomimetics or bionics. The compilation is by no means exhaustive;[107] it is only intended to give an impression of the sectors in which biomimetic innovations are present and the stage of innovation they have reached to date.

Difficulties of implementing biomimetic innovations

Many of the biomimetic developments described in the scientific and popular literature have not gone any further than the research and development stage – in some cases, for several decades – or development has been broken off. Many of the extensive biomimetics publications are therefore more likely to depict models for potential solutions to as yet unsolv-

106 In the course of this study, it was necessary to rather "look ahead" in our focus on the planned case studies. Thus, an evaluation of previous biomimetics (project) funding was neither desirable nor possible.

107 A further look at innovation activities in biomimetics reveals, for example: Bar Cohen (2006), Rossmann/Tropea (2005), Boblan/Bannasch (2004), Nachtigall (2002), Gleich (2001). Finally, in Table 12 (chapter 4.3), in which the German companies with activities in the area of biomimetics are listed, further references to various innovations and their respective stages of development are offered.

ed (or insufficiently solved) technical problems than to actually implement innovations (Bar-Cohen 2006, Küppers/Tributsch 2002; Nachtigall 2002; Greguss 1988; Zerbst 1987; Heynert 1976; Gérardin 1972). In this context, obstacles to innovation are dealt with rather rarely; at the most reference is made to the difficulties in arousing the companies' interests for these solutions or to a "conservative" engineering tradition. The attempts at an explanation,thus , tend to remain on the individual or anecdotal level.

Before attempting to explain any individual cases, however, it must be pointed out that the most important driving forces and restraints for biomimetic innovation are those that affect all types of innovations and these should first be identified.

General drivers and barriers

Innovations are principally influenced by two factors: the development of new (not only) technical capabilities (technology push), on the one hand, and a certain demand pull, on the other. Both must come together for new innovations to even arise and then succeed. Additionally, government regulation, the competitive situation in the relevant markets, and also – with respect to the potential for innovation – path dependencies and the (non)existence of a window of opportunity have an effect on failure or success (Dosi 1982; Sartorius et al. 2005; Jacob et al. 2005). That means that in all likelihood the success – or failure – of an innovation in the field of biomimetics has in many if not most cases little to do with its biomimetics origin or the nature of the innovation. In most cases, it is path dependencies that frequently hinder innovations from succeeding, even if they are clearly superior to existing products, processes, or services. The technologies a new innovation is intended to replace are frequently embedded into systems that have developed around them or been conformed to them; the accompanying systemic inertia is not so easily overcome.

Influences on innovations

Biomimetics specific barriers

Nonetheless, it certainly makes sense to look for biomimetics-specific obstacles. In light of our knowledge of the field, the interviews, workshops, and, above all, the case studies, the following hypotheses on biomimetics-specific innovation or transfer obstacles can be formulated. First of all, there are problems associated with the problematic nature of the transfer of knowledge (i.e., the biomimetic abstraction process). Thus, for example, the hope for models of resource efficiency is, on the one hand, valid, but it must be determined whether congruities exist with respect to the respective "scarcities". Ecosystems and evolution optimise in response to different shortages than industrial systems. The hope for ecological or evolutionary optimisation of biological solutions is valid in principle, but it must also be determined whether the optimisation parameters are more or less in agreement with the following aspects:

a. Safety: The individual does not count in the evolutionary process; this we cannot (and would not want to) allow in the design and safety aspects of technical systems;
b. System stability vs. system adaptability: Organisms are more or less at the mercy of an ever-changing environment; the capability of humans to design and stabilize their (technical) environments is much greater.

In the relationship between (mechanically predetermined) system dependability and adaptability, a paradigm change towards intrinsically safe, more robust and resilient technical systems may currently be underway.

In the long-term, the "lock-in" into a so far extremely successful production paradigm will also play a role. For the longest time, the industrial production paradigm was determined by the mechanistic world view and mathematical experimental science based upon it (e.g. repeatability as a criterion for success in experimental science). This could currently begin to move in the direction of a more suitable approach to "dealing with complex systems."

Finally, cultural barriers between scientific schools or between disciplinary vs. interdisciplinary approaches, between reductionism and the systemic sciences (holism) may play a role. Still, some of these theses can also be interpreted as suggesting that in this respect, biomimetics was or still is simply "too early" in some of its approaches.

While briefly observing some individual cases, these comparatively abstract theses can be illustrated. There are, for example, some suggesting reasons, why the much cited biomimetics example of "artificial shark skin" for the reduction of aerodynamic drag in aircraft construction has not yet gone beyond a few test flights some years ago. First of all, there are technical problems such as the question of how a film applied to the aircraft fuselage can successfully overcome the extreme requirements (e.g., variations in temperature ranging from -30 or even -50° C at high altitudes to more than 70° C in the sun at airports nearer the equator) without the microstructures becoming so clogged that the desired effect is lost; furthermore the inspectability of the aircraft surface with respect to the formation of cracks needs to be considered, since it is extremely important for safety reasons (e.g. cleaning, inspection for fissures and complete lacquer removal after a certain number of flight hours). Finally, the price for tax-free jet fuel is possibly still so low that the effort and expenses necessary for the attainable savings are simply not economically viable. Similar reasons can be put forward for as to why, for example, the biomimetic fuselage form developed by von Videler in 1992 for Fokker has not yet been accepted (see Biona-Report 8, 1992). In addition to its unfamiliarity of the new, manufacturing reasons

Example riblet foil

may well have played a central role; it is easier to manufacture substantially simpler and less expensive standardized tube segments and join them together and then round off the resulting front and rear than it is to work with continually varying diameters and three-dimensional inclinations. In this case, too, the current price of jet fuel may not have been sufficient to make this manufacturing disadvantage competitive.

5.3 Driving forces and obstacles in biomimetic innovation – two examples: optimisation processes and artificial photosynthesis

See chapters 3 and 4

Biomimetics: a heterogeneous field

The analyses of publications, patents, and research networks (see chapters 3 and 4) and a look at the relevant survey literature (see, for example, Bar-Cohen 2006; Nachtigall 2002) have shown that biomimetics is represented as a broad and heterogeneous field, and this is true with respect to the narrower community that identifies itself using the terms *biomimetics* and *bionics* (biomimetics in the narrow sense) as well as for the broader sense of learning from nature. Biomimetic research is carried out in a large number of disciplines, both with and without direct reference to biomimetics. This was also the basis in the choice of the case studies for the identification and assessment of current and future biomimetic trends and potentials (see chapter 3.2).

See chapter 3.2

In-depth case studies: »optimisation processes« and »artificial photosynthesis«

See chapter 3.2

Innovation processes in biomimetics

In contrast, the in-depth studies on optimisation methods and artificial photosynthesis focused more on the innovation processes in biomimetics. In addition to the trends and potentials in biomimetics, we were interested above all in the specific driving forces and obstacles in biomimetic innovation. In the case studies (chapter 3.2), our initial impression was confirmed that biomimetic R&D indeed has been going on for some time in many areas, but that successful innovations tend to be the exception – thus the question arose (especially during the in-depth studies), why innovations from the biomimetics field succeeded or failed. Two in-depth studies were carried out specifically to follow this line of questioning. For the first, the field of nature-analogue (biomimetic) optimisation processes was selected,[108] together with an approach that should focus on the techno-

108 Here, too, the finding is confirmed that biomimetic R&D often is being done under case-specific terminology. In the case of optimisation methods, these include terms such as *nature-analogue methods, genetic algorithms, particle swarm optimisation*, and *evolutionary optimization strategy*. Even the co-founder of the evolutionary optimisation strategy, Prof. Rechenberg, from TU Berlin, in the description of his Biomimetics and Evolutionary Optimisation Strategy Program, differentiates conceptually between biomimetics and the optimisation method designed by him. Nevertheless, that it deals with approaches going back to natural and biological phenomena is clearly recognizable. Genetic algorithms copy the behavior or functionality of genes; particle swarm optimisation is based on the swarm behavior of animals, for example, birds, insects, and fish; and the evolutionary optimisation strategy transfers the evolutionary development and optimisation process into technical construction processes.

logical capabilities of these processes (in the sense of "technology push") as well as one that focuses on the demand situation in selected areas of logistics (in the sense of "demand pull" as a driving force). An in-depth analysis of the technical possibilities as well as the demand situation was intended to explain why, although biomimetic solutions have been able to establish themselves comparatively well in the area of optimisation methods – at least in research and development – nonetheless comparatively few commercial success is known. The second in-depth study on artificial photosynthesis, addressed biomimetic approaches to energy conversion and its utilisation.

Technology push and demand pull

The logistics case study showed that in the area of logistics almost no biomimetics-derived approaches have so far been applied. This result is interesting because logistics represents a typical field of application for optimisation methods. Particular attention was therefore paid to this aspect in the course of the respective in-depth studies. The resulting picture is rather confusing with respect to the current relevance, the dynamics, and the future potential of biomimetic optimisation methods, which are primarily bioanalogue optimisation processes, like, e.g., the evolutionary optimisation technique (Rechenberg/Schwefel), evolutionary algorithms, swarm intelligence (the ant algorithm), and the method of computer-aided optimisation (CAO, SKO) developed by Mattheck. That may well have to do particularly with the fact that the methods as well as the respective application situations (or the respective optimisation problems to be solved) are so heterogeneous that a comparison between the bioanalogue processes as well as a comparison to non-bioanalogue competing approaches is simply not possible. The best-known bioanalogue optimisation methods are apparently in wide application, even though these applications are not always well documented. Currently neither exceptionally strong scientific dynamics nor a particular element of success can be seen in any of the bioanalogue or competing methods. Due to the continually increasing complexity in economy and, above all, in the area of logistics, in this respect rather more was expected. Stakeholders in the field of logistics explained the situation by pointing to a current lack of pressure to solve such problems (or low awareness of the need for such problem-solving). The companies are apparently able to solve their present (optimisation) problems using simpler means and are displaying a certain restraint with respect to the necessary training and adjustment period that more complex methods would require. However, this rather cautious assessment of the market success of biomimetic approaches is based on the current situation, in which the development of biomimetic methods is just beginning and current methods are just starting to bring to bear their full optimisation potential. With respect to approaches based also on Rechenberg's work –

Biomimetic optimisation processes in logistics

Biomimetic optimisation processes

Currently still weak demand for biomimetic optimisation processes in logistics

organic computing, for example – it is clear that in the area of information and communications, a comprehensive paradigm shift is imminent, one that because of the growing pressure to solve problems in the information and communications area (increasing complexity and resource input), will forge new paths. The respective systems, however, are only expected to be ready for market in the medium term.

Artificial photosynthesis

The topic of energy production and its use has also played a prominent role in the biomimetics literature for the last several decades; nonetheless, few approaches or successes in this area which go beyond the stage of basic research have been noted. Artificial photosynthesis is a research objective, where a vast number of various different approaches are being pursued, all of which are aimed at understanding and technically implementing the natural process of photosynthesis and its fundamental structures. Here, too, we rarely find the terms *biomimetics* or *bionics*, although the focus on nature's model is clearly evident.

Comparison of various biomimetic innovation processes

In contrast to optimisation methods and logistics, in the case of energy production and usage, the demand-pull factor, i.e. the need for alternatives, is much stronger than the capabilities needed to develop biomimetic-technical approaches. While there still seems to be an insufficient demand for biomimetics-based optimisation methods, when it comes to energy the technology push factor is pushing at the seams, – because the development of respective solutions (keyword: artificial photosynthesis) is taking much longer than originally anticipated. But again this should not in any way be considered as a crushing judgment against biomimetics. On the contrary – this situation is found in almost all areas of research-intensive high-tech innovation. The biotech/genetic engineering project to anchor the capability of biological nitrogen fixation in additional crop plants comes to mind; this endeavor is apparently just as far from a solution today as it was at the beginning of the 1980s, when it was first publicly proposed and marketed.

Implementation ability of biomimetic innovations

With respect to application maturity, the optimisation methods and the biomimetic conversion of energy clearly differ. Some of the nature-analogous optimisation methods could be considered well established. In the area of biomimetic energy conversion, however, almost no approaches can be identified that are already in use. Only the dye-sensitized solar cells – whose status as biomimetic technology is not without dispute – appear to be close to a market introduction. All other approaches to biomimetic energy conversion are still to be found exclusively in the state of fundamental and applied research.

Biomimetic innovations of limited importance in logistics

In both fields, the biomimetic approaches appear to occupy a subordinate position. That does not suggest, however, that they are unimportant. Especially in the case of optimisation methods, it has been shown that the biomimetic methods are finding application in areas in which conventional approaches lead either to no solutions or to too few good solutions. The field of optimisation methods is distinguished by its relatively high degree of problem-specificity. Thus, the efficiency of the various methods is generally quite difficult to compare, because optimisation problems are in part extremely variable and therefore can only be managed by very problem-specific optimisation methods. If, therefore, biomimetic optimisation methods are quantitatively not as widely in use as conventional methods, this may be because those optimisation problems which occured so far are still being resolved satisfactorily by conventional methods. The logistics study comes to the conclusion that some specific capabilities of nature-analogous optimisation methods are so far not widely in demand, but that there is some indication that this will change with the increasing complexity of logistics problems. The nature-analogue optimisation methods are effective with optimisation problems of mid- to high-level complexity. The prevailing optimisation problems in the logistics sector today evidence only a low- to mid-level of complexity, and are therefore are more easily, precisely, and quickly solved by means of mathematical analytical methods. Assuming that the complexity of logistics optimisation problems is increasing, there is reason to believe that biomimetic optimisation methods, too, will become more widespread.

Biomimetic optimisation processes of great importance in the future

We mentioned right at the beginning that the demand in the area of energy conversion is quite different from that of optimization methods. The demand for alternatives to conventional, fossil fuel–based energy technologies, however, has only recently begun to grow. Energy technologies based on regenerative energy sources presently are only able to establish themselves with the help of state support. Even though the demand is greater now than ever before, the lack of biomimetic energy conversion techniques is much stronger on the technology side, that is, in the development of realisable (stable, long-term) solutions based on the model of natural photosynthesis.

Biomimetic processes in the energy sector of increasing importance

It must be pointed out, however, that the energy supply is characterised by enormous path dependencies. The current energy system is based almost entirely on the use of fossil fuels. Technologies for conversion and utilisation of energy – for example, coal-based power plants or combustion engines in automobiles – were once (basic) innovations and have become firmly established and further developed (incremental innovation) over decades. Extensive infrastructure (filling stations, repair shops, etc.) but also institutional frameworks (emission standards, safety checks, etc.) have grown up around these technologies, leading to a solidification of the technology paths. Alternative technologies for energy conversion and utilisation, for example, hydrogen-driven unitised power stations or elec-

Energy system inhibited by strong path dependencies

Sunk costs

tric motors for automobiles, may in many respects already be superior to the established technologies. The abandonment of "old" technology pathways, however, is made more difficult by the existing infrastructure, institutional frameworks, and other societal structures due to the associated loss of investments in the existing infrastructure (sunk costs). The new, innovative technologies, on the other hand, are faced with the opposite problem: the essential support structures have not yet been established. A public supply network for hydrogen, for example, would first have to be created, which, for one thing, would require large investments. In the case of dye-sensitized solar cells, although a significant demand pull already exists, a successful implementation would be extremely difficult and prolonged even if these or other organic solar cells were in production and available on the market.

Biomimetic solar cells partly already close to the market

In order to widely succeed as a new innovation, (biomimetic) solar cells will have to demonstrate above-average benefits in comparison to existing energy attainment technologies. At the same time, the use of conventional technologies for energy attainment would have to come up against limits or obstacles that would make the abandonment of these established technology pathways more attractive or even necessary. In the case of both of these conditions, there is sufficient evidence to suggest that this will only take place in the mid- to long-term. Solar energy conversion technologies promise low-risk, more versatile and flexible applications. Biomimetic solar cells like the photosynthetic plants and bacteria, would also be capable of utilizing diffuse light efficiently and, above all, at lower temperature ranges. Due to their compact size and low weight, they could be integrated into facades, portable devices, or even textiles.

5.4 Biomimetics-specific driving forces and obstacles

Attributes of biomimetic innovations

In spite of the primacy of the general drivers and obstacles for all innovations, in our detailed survey, we were able to identify factors that specifically affect biomimetic innovation. It is a general attribute of biomimetic innovation that, in the course of research and development of the innovation, learning from nature (or from scientific results, especially in biology) is in some way involved. This means, scientific knowledge of biological events is being transferred into technological applications. This transfer process is, however, not identical in all cases of innovation in biomimetics. It differs case-by-case and can influence the entire process of innovation.

Biomimetic transfer process

The differences concern the complexity of processes and structures that serve as a model. For example the mechanisms that ants use to collectively, quickly and flexibly find the way between their nest and a food source are

Complexity of biomimetic models

relatively simple; they are based on laying down a track of volatile traces of pheromones; thereafter, some very simple rules are used to follow these traces back. The ability to solve highly complex problems (higher level) is therefore based on fairly simple principles (lower level)[109]. It was relatively easy to transfer these mechanisms of rather low complexity into respective optimisation mechanisms (ant-algorithms).

Lack of basic knowledge about biological systems

However, photosynthesis, which serves as a model for biomimetic procedures for energy generation, consists of highly complex structures and processes even on the physical chemical level; here, photon energy is absorbed, transferred, and, through a multitude of intermediate reactions, transformed into chemical energy which is then stored; therefore, emulating photosynthesis is an inordinately more difficult venture. Currently, many of the individual processes within photosynthesis are not at all or only rudimentarily understood or known; there is therefore no question of transferring them to technological processes. Accordingly, there is currently no real biomimetic-technological process that comes even close to duplicating photosynthesis. Concerning driving forces and, in this case especially, obstacles for biomimetic innovation, it can be stated that the complexity of the model and, even more specifically, the distribution of complexity in the model, severely impacts the transfer into technological processes.

Another aspect that governs the difficulty of a transfer from the natural model into a technological application lies in the ability to extract and isolate single components from the entire biological system for transfer to technological systems. For Otto Lilienthal, for example, his breakthrough in the attempt to fly like a bird was only possible after he analytically separated the lift force from the propulsion; he abandoned the emulation of the flap of the bird's wing and concentrated entirely on the lift force (gliding).

Difficulty of extracting single aspects from transfer process

With respect to the development of artificial photosynthesis, such isolation efforts are currently not in sight. It is still not clear whether the process only functions (and should only be labeled as "biomimetic") when energy is transformed from solar photons into chemically bound energy – ideally with the application of carbon dioxide and water. The structures, materials, and processes of photosynthesis seem to form a functional unit. If this unit is destroyed by removing individual parts, the entire function appears to be lost. Photons seem to be collected and transferred into the respective reaction centers only by the corresponding antennae systems. There, in turn, only certain molecules appear to be able to produce the

109 As an example of this form of dealing with complexity (not to be confused with the reduction of complexity), two considerations of Gößling-Reisemann are put forward: A research and developer today, given the assignment of developing a two-wheeled vehicle would likely not develop a bicycle. Instead of solving the drive principle in a simple manner by leaving the steering to person sitting on top of the vehicle, he would likely attempt to build in a highly complex self-steering mechanism in the two-wheeled vehicle.

From »inspiration by nature« to »copying nature«

charge carriers that are needed to split water into oxygen and hydrogen, etc. If one therefore places biomimetics innovations on a scale with "a copy of nature" at one end and "inspired by nature" at the other, one can conclude that the more tightly one is locked into the integrity or functional unity of the natural model, the more difficult the research and development aspects of the process of turning biomimetic ideas into market-ready products and processes becomes.

The specific aspect of biomimetic innovation, its reference to the model of nature, can therefore be a driving force as well as an obstacle. It is generally exactly those qualities or properties leading us to emulate a natural model that serve as driving forces in the first place. In the case of the LotusEffect®, these are the dirt-repelling properties of a surface; in the case of biomimetic optimisation methods, the ability to work even in noisy and complex environments; in the case of biomimetic means of energy generation, it is the ability to convert solar radiation into useable forms of energy, etc. When the transfer of a specific biological property into the respective technological system succeeds, it is the quality of the biomimetic solution that represents the advantage over comparable non-biomimetic systems in the innovation competition.

No anything-or-nothing-principle

However, such transfers (fortunately) do not always follow the principle of "all or nothing"; along with the desired properties and functions of the biological model comes the often more or less inevitable transfer of at least some undesirable properties. A popular example in this context is again the LotusEffect® innovation. The bump structure, which is ultimately responsible for the desired effect, also causes an optical effect, i.e. a dull surface appearance. Thus, aside from the problem of technological realisation, there will probably be no "self-cleaning" cars in the foreseeable future simply because of the prevailing preference for gleaming automobile bodies.

Example dye-based and other organic solar cells

Dye-sensitized and other organic solar cells also have certain advantages over conventional solar cells that are ultimately due to their biological model. They can utilize diffuse light, do not require difficult-to-obtain semiconductor materials, are lightweight and malleable, etc. Bio-organic materials also are generally easily biodegradable; therefore, it is perhaps no surprise that the use of organic dyes and other materials in organic solar cells is accompanied by a shorter lifetime and a comparatively rapid flattening of the performance curve over time. While plants and photosynthetic bacteria permanently regenerate the structures that are involved in photosynthesis and thus ensure a constant performance over a long period of time, organic solar cells lack the possibility of self-repair and regrowth. The fact that dye-sensitized solar cells, which are the most advanced non-conventional solar cells to date, show a maximum lifetime of only about

Artificial photosynthesis

five to ten years, while silicon-based solar cells work twenty years or more with high output, has probably blocked innovation and market introduction to a great extent.

Sum of innovation attributes counts

As with all innovations, before a biomimetic approach can push aside other, non-biomimetic solutions, it must be substantially superior in the sum of its relevant properties and functions. Thus, a negative property that cannot be isolated and is therefore also transferred can block the success of highly interesting innovations.

5.5 Prospective classification of biomimetics

Biomimetics as enabling line of technology

From the innovations' point of view biomimetics is probably no key technology that will revolutionize entire branches of industry or the national economy. It is neither a technological approach that is focused on a specific product line or sector. Biomimetics represents rather a very heterogeneous, generic, cross-sectional approach that, in principle, is applicable to many different science disciplines and areas of research and development. The most fitting characterisation is that of being an *enabling technology.*[110] Biomimetics, as an enabling line of technology, deals with a very specific history of origins and – building on this – with a very specific (solution) quality. Biomimetics, thus, is generally in competition with conventional approaches and must always prove its advantages anew on the market. As mentioned before, quite a few promising biomimetic problem-solving approaches have already run up against (so far) insurmountable obstacles at the earliest stages of innovation. Thus, there is little room for excess expectations with respect to potentially successful product launches with biomimetic solutions. The same, however, is true for broad areas of bio- and genetic technology and the material sciences. In light of existing path dependencies and systemic inertia each innovation initially faces a difficult start – and worse – when it cannot rely on existing innovation systems, and the conditions for successful innovation first need to be created. In order for innovations to move forward, substantial reasons and significant driving forces need to exist. New innovations must, for example, make it possible to open up new markets or demonstrate very far-ranging advantages as compared to current solutions, or they must be of fundamental significance for the maintenance of the competitive position of an affected firm. It is not possible to make any sweeping statements about the extent

Heterogeneity of biomimetic approaches

110 The concept of *enabling technology* is admittedly associated with a certain lack of precise definition. Used in conjunction with nanotechnology, it generally refers to the expectation that the new capabilities for design at the level of elementary building blocks – as the saying goes, "atom by atom" – open up fully new technological possibilities. This, however, is not biomimetics, which also taps into new technical possibilities, but not by means of *a technology*, but rather through a specific solution path.

to which these "good reasons" and driving forces exist in the case of many biomimetic solutions – not the least because of the strong heterogeneity of biomimetics approaches mentioned above.

Biomimetic innovations to be expected in a number of sectors

Since biomimetics represents not a line of technology, but rather a solution-finding pathway, it is in a position of influence in the most varied areas. This is especially the case in the multi-purpose technologies, such as information and communications technology, the material sciences, and nanotechnologies, in which it can provide substantial guidance for new solution-finding pathways and design options. For example, Sörensen (2004) writes with respect to IT applications, "Biomimetics is a design methodology for complex artifacts, deployment to support human design with self-organizing evolutionary mechanisms," and also points out that, for example, the IBM Autonomic Computing Project was begun against the background that the development and maintenance of IT systems of the future would only be possible through new design methodologies for such systems. Thus, it is against this backdrop that organic computing approaches are being funded in Germany, for example, in order to satisfy the prospective demand in relevant areas of application. However, a broad market penetration is not anticipated before 2015 (BMBF 2006b). This is true for a number of further areas (particularly the material sciences, proteomics, and nanotechnology), as was made clear in the case studies in particular but also in the analyses of the technology foresight studies (see chapter 3.1). A strong upturn in biomimetics approaches found in the literature and patents beginning at the latest in the mid-90s could suggest that approaches from the narrow area of biomimetics as well as more general approaches of learning from nature (keywords *self-organisation, self-healing*, etc.) could attain market relevance in a similar time frame.

See chapter 3.1

No quick solutions

The offering of new biomimetics solutions for complex problems (self-organisation processes in logistics chains, robotics, material sciences, etc.) seems to be encountering a commensurate demand, in that the complexity issue is increasingly being pushed into the foreground. However, these biomimetics avenues of research likewise do not necessarily promise rapid technical solutions. In the future, these approaches may gain substantially in significance for the various scientific and technical branches.

Biomimetics as approach towards dealing with complexity

With respect to the future role of biomimetics in the national innovation landscape, the analysis of the technology prognoses, the (in-depth) case studies, as well as the discussions in the workshops we conducted and the interviews made clear that substantial aspects of the technological direction can be tightly coupled with a few key biomimetics guiding visions. In a vast number of areas (robotics, the material sciences, information and communication technologies), biomimetic or bionic aspects play an important role. Biomimetics shows itself to be highly compatible with respect to the development of new technologies, technological developments, and market demands in various science and research areas. The increased demands on the adaptivity of technical systems and on the capability of coping with complexity have, at least in part, paved the way for the biomimetics approach. The processes of enquiry in science disciplines such as chemistry, the material sciences, nanotechnology, and information and communications technologies lead to biomimetic approaches in the broadest sense, including multifunctionality and self-healing, among others. In this sense, it is likely that biomimetic approaches will have an increasing influence on the design of the technical systems of the future.

This will especially hold true for the bottom-up nanotechnologies and their expected link to other science disciplines in the so-called converging technologies. It can be expected – and already observed – that in the various science disciplines (and their convergence) far-ranging paradigm shifts are on the way whose final successful implementation will also make necessary far-ranging adjustments to the innovation system in order that new technical economic trajectories can subsequently develop.

Adaptation processes in innovation systems required

Should this thesis prove to be true, then possibly far-ranging adjustment processes in the innovation system will at the same time be necessary, and new types of pathways will arise. The discussion about the prospects for nanotechnology and, in particular, the converging technologies make these approaches clear and are already leading to changes in the funding landscape. Biomimetics can play an essential role in this context.

5.6 Tentative conclusion with respect to biomimetics and innovation

In summary, the following conclusions can be made with respect to biomimetics-specific driving forces and obstacles for biomimetics innovations:

(Non-) equivalence between the complexity of the natural model and the biomimetic solution (complexity-match/complexity-mismatch)

The analytic and technical isolability of certain subfunctions (or steps) in the natural model makes the transferability easier (for example, the severability of lift and propulsion in the case of the aircraft wing or the severability of the desired form from the natural material). The analytic and technical non-isolability of subfunctions, processes, or structures makes the transferability more difficult and increases the need for a complex "replication" (for example, the complexity of natural photosynthesis).

Here, one must also include the still unsolved technical problems of the replication of hierarchical materials, the realisation of a "genuine" lotus surface, or the realisation of self-organising and self-healing (repairing) technical systems. Finally, it was found that a more dynamic demand for natural-analogue optimisation processes may only be imminent, when the technical and, for example, logistic problems require higher capabilities for coping with complexity.

If one therefore places biomimetics innovations on a scale with "copy of nature" at one end and "inspired by nature" at the other, one can conclude that the more tightly one is tied to the natural model, the more difficult the research and development process of turning biomimetic ideas into market-ready products and processes will be. However, if one can strongly abstract from the natural model and transfer only single aspects set free from the systemic context, the implementation process will be easier and probably also more rapid.

The interrelationship between desirable and undesirable functions, structures, and properties (functional trade-offs)

If preference is given to biodegradable materials and supplies, the problem of the limited stability of these substances arises. Thus, solutions need to be found that allow substances to be kept stable in some environments and biodegradable in others; nature does this, for example, with wood, which remains stable in dry conditions and breaks down in conditions of moisture. Another example is found in dye-sensitized solar cells, whose life span is presently rather limited due to the organic materials used; so far, the many advantages of dye-sensitized solar cells associated with the deployment of organic absorbers (greater flexibility, utilization of diffuse light sources, lower toxicity) still seem to be inseparably linked to the ever-present aging processes of nature. A further example of such trade-offs would be the (genuine) lotus effect, which (unlike merely hydrophobic surfaces) is not realizable as a gleaming surface.

Biomimetics as an orientation and implicit model for future technologies

In general it is also clear that the biological model increasingly provides an orientation function for technological development. As a consequence, biological systems are increasingly being analyzed in an interdisciplinary manner in order to identify technological problem-solving approaches. The information and communications technologies, in particular, make possible new ways of approaching analysis and modeling. In the case of nanotechnology, reference is increasingly being made to bionic or biomimetic approaches, whereby one of the fundamental elements of nanotechnology, namely the bottom-up approach, makes reference principally to the natural "production processes."

The important driving forces, which also could mean a push, above all, for biomimetics in the broad sense (learning from nature) and the third strand of nanobiomimetics, may essentially resemble those that determine the development dynamic of the material sciences and nanotechnologies, as well as the hardware and software in the information and communications technologies. The basic idea of the bottom-up approach, the option of self-organisation or self-growth, opens up interesting new prospects for nanotechnology and biomimetics alike. However, the anticipated push, the expected dynamic still tends to be technology driven. The demand side, however, may go on to play an ever greater role in a series of fields of application where the borders of conventional technological approaches increasingly turn into stumbling blocks, thus intensifying the search for new technological options.

Biomimetics as an expansion of the field of inquiry

In areas in which obvious obstacles to generating solutions on the basis of existing approaches or technologies exist, the biomimetics approach seems to rank highly, inasmuch as it expands the problem-solving realm (analogy formation).

Biological systems demonstrate properties and capabilities that to some extent are unattainable in current technical systems. Should the need arise (or exist) for high-performance technical systems with properties and capabilities similar to meeting or even exceeding those of their analogue biological systems, biomimetics innovations could step in to meet the demand.

Biomimetics – a positive image for new technologies

Biomimetics enjoys an extremely positive image among the public. The many illustrated books that have appeared in the past several years (Blüchel/Malik 2006; Nachtigall/Blüchel 2003; SiemensForum 1999; WWF 1991) and (popular) science books (Blüchel 2006; Cerman et al. 2005; Kesel 2005; Willis 1997; Marguerre 1991) as well as a series of television and radio programs on biomimetics clearly demonstrate the positive image that biomimetics carries. Due to fascination which derives from the remarkable performance of natural organisms and their structures and processes, biomimetics is viewed as a modern high-tech affair. At the same time, biomimetics carries no negative associations, because it is nature, after all – which generally is viewed positively – that serves as the biomimetics model.

Various levels of innovation success

Viewed as a whole, one can discern various levels:

Successful innovation depending on degree of biomimetic abstraction

Importance of knowledge about exact functioning of natural model

On the one hand, it would seem that the innovative capability or success of biomimetic approaches is at times dependent on the specific point on the above-mentioned continuum of nature at which actual development begins. Thus, in most of the successful biomimetics innovations, for example, the winglets on the wingtips of aircraft or (in their current form) the lotus-effect surfaces, even though the exact functionality of the natural model may not be precisely known or the exact causal relationship fully understood, emulation of the natural model was nonetheless possible; a particularly high degree of abstraction was still possible, and the challenge of sub-aspects, for example, the transfer of the form but not the material, did not lead to a failure of the desired effect.

In the case of artificial photosynthesis, however, the situation is clearly different. Here, biomimetic applications are still a long way off, since an emulation or duplication of the natural process, namely the conversion of photon energy into chemically bound energy can only succeed by means of similarly complex structures at the molecular level; this furthermore requires a certain insight into the chemical and physical processes in plants or bacteria, which will still require some years (even decades according to some experts) of fundamental research.

Successful innovation depending on competing trajectories

Path dependencies

In the case of individual biomimetic developments, one must consider the competitive position of the biomimetics solution on the market should its development proceed to the stage of being ready for market introduction. In the implementation of technological development trends, biomimetic solutions are by no means given preferential treatment. As a rule, it is much more likely that existing technological approaches will be further developed, thus consolidating and strengthening existing developmental trajectories (see the example of optimisation methods above). Alternate technological approaches only succeed when the existing line of technology comes up against insurmountable obstacles or when the operational capability of the alternative (biomimetic) solution can be increased so that the implementation is economically viable. This becomes a problem, in particular, when the need for new technological approaches is fundamentally necessary. As an example natural sensors currently far exceed the capabilities of artificial sensors; nevertheless, in the end it is the costs which impede the development and, above all, the implementation of biomimetic sensors.

Technology race

To make matters worse, a certain technological competition arises in which traditional approaches and methodologies are steadily improved, thus achieving the functions and qualities that the biomimetic solutions can principally offer but without the necessity of (explicitly) having to rely on a biomimetic solution. The growth of hybrid (technological) developments can also be assessed to some extent; these developments further expand on and refine conventional technologies, thus implementing some aspects of biomimetics (smart technology).

Biomimetics as scientific field

Finally, as a general rule, the development of technologies and particularly their commercial success does not depend on the respective processes or methods being implemented, but rather on the respective functionalities; the actual manner by which this functionality is attained is of little interest to most end users. On the market little attention is paid acknowledging the role of biomimetic problem-solving approaches.

In addition to the already mentioned driving forces and obstacles found along the path of the innovation process, there is a series of other factors that affects the field of biomimetics as a whole:

Biomimetics is (still) characterized and carried out by a small community and individuals

The explicit biomimetics community (biomimetics in the narrow sense) is rather small (see chapter 4 and its impact on the scientific and enterprise communities is inevitably limited. Nevertheless, it does succeed to a certain extent in making its presence known.) To increase and advance the biomimetics approach, it is therefore necessary to (further) develop the community base (as is already occurring by means of adequate funding approaches). The promotion and funding of nanotechnology has clearly shown how a field can be developed under such an "umbrella."

See chapter 4

Biomimetics as a field has at its disposal only limited institutional resources

Till this day, biomimetics research in an institutionalized form is neither conducted by biologists nor engineers to any significant extent; those engaged in biomimetics are not being promoted by any educational, research, or development system. The few academic and non-academic institutions, such as courses of study and lectures or R&D departments, societies, and congresses (see Nachtigall 2002, 19 et seq./35 et seq. and chapter 4.7 of this study for an overview), have not been able to change much so far. Therefore most of the innovation efforts in biomimetics to date can be ascribed to the strong commitment of individual and small groups. Of course, this is also true for other forms of innovation, but biomimetics seems to be a special case.

See chapter 4.7

The formation of research networks is in its initial stages, but is essential for the success of biomimetics. The findings on innovation networks and systems in general support this, as do the specific nature of research and development in biomimetics (but also more generally the "general purpose technologies": broad areas of deployment, limited access to key knowledge about potential fields of deployment). Their basis is namely the propagation – i.e., transfer – of biological knowledge and findings into genuinely non-biological contexts. Therefore the technology and knowledge transfer, which has also been identified as an important determinant of innovation processes in other contexts (see Walter 2005), has with respect to biomimetics an exceptionally pre-eminent status (Zerbst 1987, 22 et seq). This hypothesis was also confirmed numerous times in the interviews.

Biomimetics and business

Companies are presently only able to pursue and implement research projects in the field of biomimetics to a limited extent; the analyses made clear that openness with respect to biomimetics approaches is very much dependent upon a few individuals within a company. Biomimetics appears generally to be still very much limited to areas of academic and other non-private forms of research and development (see chapter 4). When partners cannot be found in private enterprise, biomimetics researchers frequently take on the transfer of biomimetics approaches into products and processes themselves (hive-offs). This was determined in an analysis of the biomimetics innovations found in the comprehensive biomimetics literature (for example, Nachtigall 2002) and was also confirmed in interviews. The lotus effect is exactly such a case, in which it became finally necessary for the biologists to become entrepreneurs.

See chapter 4

Traditional disciplinary approaches hinder the implementation of biomimetic approaches

Biomimetics necessitates a particular degree of interdisciplinary work. Biological results need to be developed against the backdrop of present and possibly future technical demands and then transferred into technical applications. This process requires a very high degree of interdisciplinary skill on the part of the players involved, which must be implemented into priority interdisciplinary projects and networks of biomimetics research and development (see Specht 2005, 252 et seq).

Nevertheless, education, research, and development are still very strongly organized according to discipline; interdisciplinarity is merely put into practice in neighboring or relatively closely related disciplines. Biomimetics, however, connects science disciplines that have traditionally always been widely separated from one another – and today still are. Therefore, very few individuals presently fulfill the requirements of biological-engineering interdisciplinarity. The methodologies, terminology, and also the scientific images of both disciplines are very different, which greatly complicates a coming together and communication on both sides.

At those junction points where the capabilities of interdisciplinary research and development do exist, there is often a further lack of opportunities to put these capabilities to use (statements from interviews), because (public) research funding is also very strongly aligned by discipline; therefore, the acquisition of research funding for biomimetics projects is almost impossible. The most recent biomimetics funding has brought about important positive changes in this respect.

Fascination with biomimetics at the idea level – skepticism with respect to implementation

Biomimetics is still far removed from the scientific and industrial mainstream. Biomimetics, as a fusion of biology, hard science and engineering, is perceived in many areas of research and development – and also in business – as exotic; admittedly, here and there it may have stumbled across some interesting innovations, but above and beyond that, it does not really promise much in the way of systematic and widespread generation of successful innovation. And that which sparks fascination and enthusiasm in the public eye often evokes skepticism among researchers and businesses ("Will this really work?"). The borders between (biological) disciplines and biomimetics (and/or biomimetics approaches) are still as strong as ever: even established representatives of biomimetics refer to themselves by their (traditional) professional discipline. One reason for the lack of acceptance of biomimetics approaches among established scientists and developers may also be the lack of a resilient methodology for working in the field of biomimetics, which could serve as valid, straightforward proof of the systemic viability of learning from nature.

Customizability of biomimetics solutions frequently limited

In biomimetics it is the superlative performance of biological systems that provides the impetus and model for technological innovations. Most of the time, however, a complete copy of the biological system is neither possible nor desired, and as a rule only a few properties, structures, and functions or principles of the biological model are carried over into the technical system being developed. Thus, other components of the technical system are not based on biomimetics developments and therefore show fewer advantageous features. But in order to succeed on the market, new innovations – and not only those from biomimetics – must be superior in the sum of their properties to existing products in order to experience a demand large enough. In this respect then, biomimetics innovations fail when – although clearly superior with regard to some characteristics – they achieve only average or even below-average values with regard to other characteristics.

Material-structure-function-units of biological systems

Furthermore, biological systems form in many respects material-structure-function units; these are so strongly connected that a specific function at the same level of quality can only be attained in conjunction with the corresponding materials and structures. For biomimetics and biomimetics innovations, this means that certain high-performance technical systems based on the model of nature frequently must allow for compromises with respect to other aspects, which may consequently make an innovation unattractive. For example, biological materials are in many respects superior to technical materials (multifunctionality, weight-volume-performance relationship, environmental compatibility, etc.), but unlike technical materials, they tend to age very quickly. In biological systems this is compensated by permanent regrowth, which in the case of technical systems is currently – and likely for some time in the future – not possible.

6 CONCLUSION

»Technology as the delivery and mediation of the creations slumbering the womb of nature, this belongs to the most concrete aspect of concrete utopia.«

ERNST BLOCH, 1885–1977

6 CONCLUSION

6.1 Trends in biomimetics

Three main strands of biomimetics

At a greatly simplified level it is possible to distinguish three main strands of biomimetic research and development with respect to the time period and approach, whereby the contribution of the third strand consists largely of efforts to overcome the limitations in the two preceding strands. The three strands can be characterized as: a) functional morphology – form and function; b) biocybernetics – information processing, sensorics (sensor technology), and robotics; and c) nanobiomimetics – molecular self-organisation and nanotechnology.

Functional morphology

First and oldest development strand: functional morphology – form and function

The first and oldest of the three developmental strands focuses on the relationship between biological forms or structures and their functions. Among the most successful innovations in this strand to date are the parachute, the lift-generating aircraft wing, the streamline form, and the hook-and-loop fastener. As long as scientific observations of nature remained in the macroscopic realm, technical implementations within this same dimension using the correspondingly available techniques and materials were particularly successful, but only if the function of interest had more to do with the form and less the form-giving material. For the lift-generating function of an aircraft wing, e.g., it is the form, above all, that is decisive; its technical realisation in a non-biological material changes little. It is interesting to note that many of the examples of success in this form-function strand of development are derived from the field of fluid dynamics. This leads to a second condition for success: part of the success of biomimetics in the area of fluid dynamics is due to the fact that the biomimetic approach was (and partly still is) capable of compensating for the limitations of the analytical approach of mathematical experimental physics. Neither the analytic nor the newer numerical models of fluid mechanics were capable of making calculations or predictions precisely enough to be able to work out optimisations on the chalk board, so to speak. In the end, it was always necessary to carry out an empirical experimental trial optimisation process in the fluid-dynamics test chamber – and in such trial and error optimisation, biological evolution has an enormous lead.

Implementation problems related to production

However, the further that research on the relationship between structure and form moved from the macroscopic to the microscopic and then on into the nano-realm, the greater the technical implementation difficulties or "manufacturing issues" arose; they are among the most significant limitations today on far-reaching innovation in this field. Many highly interesting discoveries, for example, in the areas of surface structure and functionalized interfaces in biological systems (such as shark skin/riblet film, the lotus effect, and anti-fouling) have not yet been technically implemented for production so that the quality of the result would be comparable to the corresponding natural sources of inspiration. In these examples it is the quality of the technical implementation that is decisive for the desired functionality and thus for the success of the innovation.

Change of production paradigm

For the production of micro- or nanostructured surfaces and even more so for the manufacture of hierarchically structured biological materials based on models from nature – for example, bones, teeth, nacre or spider silk – a fundamental change in our production paradigm is unavoidable. In the first production paradigm a form was cut out of a given material (for example, stone or wood); in the succeeding paradigm the raw material (for example, metal or concrete) was first converted to a homogeneous mixture and then poured or pressed into a form. But it is not possible to create hierarchically structured materials in this manner; a third production paradigm is necessary: "letting things grow." It may well be that only with the help of self-organisational processes in combination with control of surrounding conditions (context control) we will be able to manufacture hierarchically structured materials; biomimetic learning can no longer focus simply on the form, but must also look to the biological process of its genesis, i.e., biological development and growth processes. It is in the third strand of biomimetics development via the bottom-up technologies that such technological prospects could become possible.

Signal and information processing, biocybernetics, sensorics, and robotics

Second main development strand: signal and information processing, biocybernetics, sensorics and robotics

Whereas the first strand of biomimetics development depended on the relationship between form and function, it is the cybernetic control loop that is characteristic of the next strand. As in the field of technical biology, models borrowed from the fields of engineering and technology provided the initial inspiration for explaining "biological" phenomena. The fundamental approaches and models of biocybernetics, sensory- and neurophysiology, as well as ecosystem theory, were initially developed in technical areas such as electrical engineering (resonant circuits, feedback effects, and control circuits, as well as sensors, and actuators), far from the field of biological phenomena. Only with the help of these models could important progress in biology – particularly in biocybernetics, sensory physiology, and neurophysiology, and even brain research – be achieved; this in turn positively influenced (not only biomimetic) technical developments in sensorics, information

Biomimetics and the heritage of artificial intelligence

processing, and robotics. Ultimately, it is the human mind and body that serve as the still unattainable model, above all in many areas of sensor technology, robotics, and information processing, up to and including artificial intelligence (AI). In the years following the initial euphoria, however, the area of artificial intelligence has become noticeably quieter. But if – for good reason – we include current approaches such as decentralized control, neural networks, and organic computing among the biomimetic solutions based on models from nature, there once again seems to be a new era dawning in this second strand of biomimetics development, because – particularly with the aid of the biomimetic approaches to managing complexity – some of the meanwhile accumulated limitations in the area of signal and information processing and robotics can be overcome. This second, rather more biocybernetic strand of biomimetics seems to be inheriting the legacy of artificial intelligence and picking up speed via the fusion of robotics, sensor technology, and prosthetics.

Nanobiomimetics –nanotechnology and molecular self-organisation

Third and most recent strand of development: nanobiomimetics – molecular self-organisation and nanotechnologies

The third and most recent strand of development in biomimetics is found at the nano-molecular and level. This strand, too, has quite a long history. Carried along not least by the push forward in nanotechnologies as a whole, developments in this strand only now seem to be reaching a breakthrough (see, for example, spider silk, biomineralisation, functionalised surfaces, template-controlled crystallisation, neurobionics and nanobiomimetics, etc.). The nanobiomimetic strand focuses on processes of growth, ontogenetic development of molecules, cells, and tissue, including their reconfiguration (reaction to load) and (self-)healing. In this third, at the moment extremely dynamic line, we begin to find some very promising approaches to solutions, even for the earlier limitations of the first and second strands of development mainly due to "manufacturing issues". Such promising, new approaches may also lead to methods for manufacturing hierarchically structured anisotropic materials based on the models of bones, teeth, or plant stem. Taking a more distant perspective on development, we can expect to find "smart materials" capable of reacting to differing loads and, if necessary, even repairing themselves. Presently – and for the foreseeable future – the strongest dynamic among the three strands is likely to be found in this rather developmental biology–oriented strand of biomimetics (learning from ontogenetic development processes), both with respect to the dynamics of the research itself as well as the possibilities for implementation.

Convergence of the development strands

Current trend: convergence of three main strands of development

Finally, a trend towards convergence of the three most important biomimetic development strands can be observed. This convergence has its complement in a broader convergence movement in exactly those advanced technologies in which learning from nature is among the most advanced research and development approaches. They include the information and communication technologies, nanotechnology, biotechnology, the material sciences, robotics, and the cognitive sciences. This is a prominent topic of discussion under the heading "converging technologies." It is likely that biomimetic approaches will play an essential role in the course of this consolidation process; in all of the areas mentioned, the importance of the concepts of self-organisation, adaptivity, self-healing, and self-optimisation rank very highly.

6.2 Technology potential of learning from nature

High importance of approaches to learning from nature in technolgoy processes

An investigation of the significance of biomimetic approaches in selected fore-sight studies confirmed and differentiated the evidence on biomimetics trends. Certainly, the topic in these studies was not biomimetics itself, but much more the focus on the model taken from nature, i.e., "learning from nature." Models derived from nature play an important role in the following fields: i) the material sciences and nanotechnologies (biomaterials; complex polymers; intelligent, adaptive, and self-healing materials; self-assembly), ii) medical technology (prosthetics, biocompatible materials, tissue cultivation, clinical diagnostics), iii) robotics and locomotion, as well as iv) sensorics and information processing (nano- and biosensors, bioelectronics, biocomputers, network topology, fault-tolerant and self-repairing systems, neural networks, organic and autonomic computing, evolutionary algorithms, artificial intelligence). In two studies explicit reference was also made to technology convergences.[111] The multiple references to the problem of "managing complexity" should also be mentioned. It is singled out as a problem of increasing importance, along with the hope that biomimetic approaches will be able to make a substantial contribution in this area.

Result of case studies: biomimetic approaches widely spread

Based on the results of the analyses of the technology foresight studies, case studies were carried out on the following fields of science: the material sciences, information technology, robotics, sensorics, prosthetics and

111 Key Technologies Expert Group (2005): Creative System Disruption. Towards a Research Strategy Beyond Lisbon. Draft Final Report; Priami, Corrado (2005): Complexity and Systemics. Expert Report for the Expert Group "Key Technologies for Europe."
In some of the subfields analyzed, such as IT (Microsoft) and nanotechnology (Roco), but also in the developmental orientations of space travel (NASA), biomimetics and bionics are expected to play a central role in the long-term.

tissue cultivation, nanobio-, and converging technologies. Much as in the foresight studies, explicit references to biomimetic terms are not found in all of the fields investigated; however, in all of the case studies we found references to the potential of learning from nature and from principles of evolution. The latter is mostly in reference to the challenges involved in dealing with adaptive, self-organizing, smart or generally intelligent applications.[112] Terms on sensorics in the general technical literature that can be interpreted as concepts belonging to biomimetics and bionics include *intelligent, smart* (Liu et al. 2005), *learning* (König et al. 2001, among others), *self-calibrating, self-monitoring, self-optimizing, self-testing,* and *self-sufficient* (Mackensen/Kuntz 2004, among others). The requirements of smell and taste sensor technology, sonar, electrolocation, strain sensors and infrared detectors are clearly oriented towards the human or animal model. Very closely related are robotics' objectives when the goal is to create systems that can fulfill several of the functions characteristic of living organisms, particularly the independent (forward) movement in space, the perception of the surroundings, and the response to environmental stimuli as well as (depending on the situation) the goal-oriented performance of specific actions. In the area of information and communication technology hardware and software, the focus on natural models is primarily a matter of robustness and adaptability (self-repair) as well as finding suitable ways of dealing with complexity (evolutionary algorithms, neural networks, organic and autonomous computing). The goal is IT systems possessing "self-x properties": self-configuring, self-optimizing, self-healing, and self-protecting.

Learning from nature and self-x-attributes prevailing – »bionics« / »biomimetics« rarely used

In addition to sensorics, robotics, and information and communication technology, the developments in the material sciences show an increasing affinity to biological models. The biomimetic idea of learning from nature entered the material sciences in the last decade of the 20th century parallel to but also as a part of nanotechnology and computational material science. Ultimately, it was the fascination derived from not-yet-achieved capabilities and combinations of properties in biological materials (anisotropic, hierarchically structured, intelligent, smart, adaptive, self-healing) that led to this focus on natural models. Particularly important here was the newfound opportunity, by means of developments in bottom-up nanotechnology, to learn engineering and technological ways of dealing with cross-scale complexities and to be able to truly manufacture biomimetic materials. An important common ground of the material sciences, nanotechnology, and biomimetics lies in their inherent tendencies toward a convergence of lines of technologies that were previously separate (converging technologies). There is some evidence that the reason for this trend to-

Especially materials sciences have increasing affinity to biomimetic approaches

112 For example, the hierarchical structuring of organic materials and substances.

Converging technologies

wards convergence is not (only) a logical consequence of the respective technological development trajectories (technological push) but rather a reaction to the ever more complex demands arising from the real-world implementation of their capabilities (technology pull). In this light, one can look at the material sciences, nanotechnology, and biomimetics, with their focus on the management of complexity, as being particularly "modern" lines of technological development. Bionics and Biomimetics play a double role with respect to the converging technologies: They are an essential element of numerous problem-solving approaches, and, as in the material sciences and nanotechnologies, they serve as a model and experimental field for such convergence.

Difficulties of separating biomimetics from non-biomimetics

This convergence is logically associated with increasing difficulties in delimitation; it becomes more and more difficult to determine whether we are actually dealing with biomimetic research approaches or merely biotechnological approaches with bionics or merely IT, sensoric or robotic approaches. With the example of spider silk this can be exemplified. A fascination with the properties of spider silk and how it is "manufactured" in the organism of the spider is shared by all alike. For exponents of biomimetics, it stands as a guiding principle or model how appropriate technology and production should look like. With respect to an appropriate biomimetic manufacturing process for spider silk, opinion may differ considerably. Implanting the gene for the production of spider silk in goats so that they would produce the spider silk in their milk would clearly be going too far for some biomimetic researchers. The genetically engineered production of spider silk by the bacterium E. coli would also be met with scepticism on behalf of many.

Example artificial spider silk

Higher depth of intervention especially with nanotechnical and gene-modified applications

However, the question of appropriate quality, possible risks, side effects, and the consequences of specific manufacturing processes is not unique to the example of spider silk. It is closely related to the potentialities of molecular design in the nanoscale realm and even more closely related to the convergence between nanotechnology, biotechnology and genetic engineering. The drastically increased depth of intervention in biological systems leads to an increase in formative power just as drastic; these are obviously associated with greater risk potentials and with an increased probability of far-ranging undesirable (and unforeseen) side effects and consequences (for nanotechnology, see Steinfeldt et al. 2007). Particularly in the area of nanobiotechnology and a possible nanobiomimetics, the objectives often go far beyond the imitation of nature: nature does not simply serve as a model for technological development; rather a three-stage process follows: (i) analysis of nature (molecular machines), (ii) analysis of the technical requirements of adaptation for industrial processes, and (iii) adaptation of (molecular machines of) nature to these production-

Technomimetics

technical specifications. In this sense, there is a transition from the biomimetic (technology modeled on nature) to "technomimetics" (biological systems modeled on engineering and technology). We can thus ascertain two orientations with presumably very different consequences with respect to their associated application perspectives and risk dimensions: the biomimetic, with biological solutions as its models, and the technomimetic, which delves into the plasticity of biological systems and then attempts to mechanize them (see, for example, approaches in synthetic and systemic biology). In both cases, the further technical solutions digress from the model of nature or natural points of origin, the weaker the aspect of being "tried and proven by evolution."

6.3 Biomimetic topics, stakeholders, and research networks – results of the literature and patent analyses

German biomimetics community leading in networking actors

Germany is leading the way in networking biomimetic research. From the annual figures for publications and patent registrations, one can see that government funded network support has increased the visibility of the activities and promoted the exchange of information among researchers and, in part, businesses. With the exception of Great Britain, such institutional research networks do not yet exist in other countries; thus, the rather small circle of explicit biomimetic programs and activities has at its unique disposal a vehicle for biomimetic cooperation and partnerships, making possible the dissemination of ideas as well as opportunities to establish contact with researchers whose work may involve biomimetic approaches that have not yet been classified as such. The next step is the integration of potential biomimetic stakeholders working in the broad sense of biomimetics. The increase in network members suggests that this is beginning to succeed.

Few actors in Germany explicitly involved in biomimetics

Significant increase of patent releases related to biomimetics – in Germany and worldwide

In order to gain insights into the current research activities and status of biomimetic approaches in various fields of science and countries, an analysis of the scientific literature and a patent survey were carried out. For the period from 1990 onwards, a clear increase in publications having a direct connection to biomimetics (search terms *bionic** and *biomim**) can be seen; the same is true for patents from 1995 onwards. Biomimetics plays a role in a diverse number of science disciplines, particularly in medicine, chemistry, biochemistry, and the material sciences, with most patents being recorded in the fields of chemistry and medicine. However, in an international comparison with respect to biomimetic research in the narrow sense, no particular strengths could be identified in German biomimetics on the basis of the literature and patent analyses. It was likewise not possible to identify any particular areas of research emphasis in German biomi-

German biomimetics R&D not particularly strong

metics. This suggests that those biomimetics stakeholders in Germany who use the terms bionics and biomimetics are certainly well networked, but do not yet sufficiently transfer this networking to international publishing activities or patents.

An inquiry using search terms for biomimetics in the broad sense (*self repair*, self heal*, self organi*, as well as biolog* + inspir**, and *bioinspir**) produced most published articles in the disciplines of physics, chemistry, computer science, and the material sciences. German research institutions are well represented in these areas and surpass Japan, France, and Great Britain. Companies from the USA and Japan are particularly active with respect to patent applications. Among German companies, the information and communications field stands out in particular, especially Siemens and Infineon.

German actors pursuing »learning from nature« rather weakly networked

However, the network analysis carried out for biomimetics in the broad sense suggests that networking of the German research landscape is below average when compared internationally. The reason may be that explicit network funding focuses on those biomimetics stakeholders gathered under the umbrella of biomimetics in the narrow sense; thus they may not be adequately reaching those researchers falling into the category of biomimetics in the broad sense. The actual reasons, however, cannot be derived from the data.

6

Germany strongly represented in section »learning from nature«

Disregarding the degree of networking and considering the more general approach of learning from nature, Germany is quite strong. It is well represented in publications, but with a smaller impact (frequency of citation – particularly in comparison to contributions from Great Britain or France), and a much smaller level of patent activity in comparison to Japan puts this into perspective.

Particular importance of biomimetics for scientific and technical education

With respect to the future development of biomimetics, it is particularly important to look at education and training. Biomimetics possesses an enormous potential to inspire in young people an interest in the natural sciences and technology. This is of special importance now – although the natural sciences are not particularly in demand in school, employees with a good science and engineering education are very much needed; employees with science and engineering training are a prerequisite if companies in the industrialized countries are to compete on the international market. The present and likely even greater lack in the future of scientists and engineers could be successfully counteracted with the help of biomimetics content in the curriculum – in the primary as well as secondary schools. Members of the (narrow) biomimetics community are already taking advantage of these opportunities and should be supported in this.

Biomimetics at universtities: rather weakly represented

The situation in the area of higher education, on the other hand, is more difficult to assess: only a small number of universities presently offer explicit biomimetics courses and lectures – which does not necessarily rule out the possibility that many more professors are dedicating resources to biomimetic topics. Particular challenges that need to be overcome in the further integration of biomimetics in education are interdisciplinarity and, in particular, proximity to the converging technologies. A tremendous future potential is generally ascribed to biomimetics (in the broad sense). To further its establishment in the universities, experts recommend the introduction of postgraduate courses of study, project support, and the establishment of interdisciplinary research laboratories. The establishment of full-fledged biomimetics degree programmes appears to be the less preferable alternative, according to the experts.

6.4 Biomimetics and innovation

Many biomimetics-related innovations still in R&D stadium

Many biomimetic research and development projects do not make it beyond the R&D or prototype stage, and among the successful innovations are few far-ranging innovations capable of opening up new markets; rather, one finds innovations consisting generally of incremental improvements.

General innovation drivers and barriers have impact on biomimetic innovations too

Before searching for biomimetics-specific driving forces and obstacles, it must first be noted that the most important drivers and obstacles for innovation in biomimetics are the same ones that affect innovation in all areas. Innovation is fundamentally influenced by two factors: the development of new (not only technical) capabilities (technology push) on the one hand, and a form of demand pull on the other hand. Both factors must come together to make it possible for new innovations even to arise and then succeed. Additionally, government regulation and the competitive situation in the relevant markets have an effect, and also – with a view to the potential for innovation – path dependencies have a slow down and windows of opportunity an accelerating effect. That means that with all likelihood, the success – or failure – of an innovation in the field of biomimetics has in general little to do with its biomimetic origin or the nature of the innovation itself.

Path dependencies

In most cases, it is path dependencies that lead to innovations often not succeeding, even when they are clearly superior to existing products, processes, or services. The technologies that a new innovation is intended to replace are frequently embedded in systems that have developed around them or been conformed to them; the accompanying systemic inertia is not so easily overcome.

Biomimetic specific innovation drivers and barriers

Nonetheless, it certainly makes sense to look for biomimetics-specific obstacles to innovation. In light of our knowledge about the field and from interviews, workshops, and, above all, the case studies, the following hypotheses with respect to biomimetics-specific innovation and transfer obstacles can be formulated. First, there are the difficulties associated with the problematic nature of the transfer process itself (i.e., the biomimetic abstraction process). For example, the prospect of models of resource efficiency is valid, yet it must be determined whether congruities exist with respect to the corresponding "scarcities." Ecosystems and evolution possibly optimize with respect to other shortages than do industrial systems. The prospect of ecological and evolutionary solutions of being optimized also has its validity. But it must be determined whether the optimisation parameters are more or less in agreement with respect (for example) to: a) safety – the individual does not count in the evolutionary process; this we cannot (and would not want to) allow in the design and safety aspects of technical systems; b) system stability vs. adaptability – organisms are more or less at the mercy of an ever-changing environment; the capabilities of humans to design and stabilize their (technical) environments is much greater. In the relationship between (mechanically predetermined) system stability and adaptability, a paradigm change in the direction of intrinsically safe, more robust, and resilient technical systems may currently be underway. In the long-term, the "lock-in" in a particularly successful paradigm of production will also play a role. For the longest time, the industrial production paradigm was determined by the mechanistic world view and a mathematical experimental natural science based on it (for example, repeatability as a criterion for success in experimental science). This could begin to move in the direction of a more suitable approach to "dealing with complex systems." Finally, cultural barriers between scientific schools, between disciplinary vs. interdisciplinary approaches, and between reductionism and the systemic sciences (holism) may play a role. Still, some of these theses can also be interpreted as suggesting that in this respect, biomimetics was or is "too early" in some of its approaches so far.

Biomimetic transfer process

Differing optimisation parameters of biological and technical systems

Lock-in

Dealing with complex systems

In the case studies on optimisation processes and logistics, as well as artificial photosynthesis, we explicitly addressed the question of driving forces and obstacles in innovation, with the following results:

Results of in-depth case studies

The best-known bioanalogue optimisation methods are apparently in wide application, even though these applications are in part not always well documented. Currently neither an exceptionally strong scientific dynamic nor a particular element of success can be seen in any of the bioanalogue or competing optimisation methods. Due to the continually increasing complexity in economics and especially in the area of logistics, more was really expected in this context. Those biomimetics actors involved in the field of logistics explained the situation by pointing to a current lack of pressure to solve such problems (or a too-low awareness of the need for such problem-solving). Companies are apparently able to solve their present (optimisation) problems using simpler means and are displaying a certain res-

Relatively broad use of bioanalogous and biomimetic optimisation processes

Rather weak demand for biomimetic optimisation processes in logistics

traint with respect to the necessary training and adjustment period that more complex methods would require. However, this rather cautious assessment of the market success of biomimetic approaches is based on the current situation, in which the biomimetic methods are developing slowly and competing conventional methods are just beginning to bring to bear their full optimisation potential. With respect to approaches based also on Rechenberg's work, for example, organic computing, it is also clear that in the area of information and communications a comprehensive paradigm shift is coming, one that – not least because of the growing pressure to solve problems in the information and communications area (increasing complexity and effort) – will forge new paths. The respective systems, however, are only first expected to be ready for market in the mid-term.

Growing complexity of technical problems increases importance of biomimetic optimisation processes

The topic of energy production (transformation) and use has also played a prominent role in the biomimetics literature for the last several decades; nonetheless, few approaches or successes in this area have been noted that have gone beyond the stages of basic research. In the case of energy production and use – in contrast to the relationship of optimisation processes and logistics – the demand pull factor is very much stronger than the capabilities needed for the development of biomimetic-technical problem-solving approaches. While there still seems to be an insufficient demand for biomimetics-based optimisation methods, when it comes to energy, the technology push factor is in fact pushing at the seams: the development of the respective solutions (keyword: artificial photosynthesis) is simply taking too long – much longer than was originally anticipated. But this, too, should not in any way be considered as a crushing judgment against biomimetics. On the contrary – this situation is found in almost all areas of research-intensive, high-tech innovation. The supply of energy is characterized by immense pathway dependencies. The biotech/genetic engineering project to implant the capability of biological nitrogen fixation in additional crop plants comes to mind; in this endeavor, research is apparently just as far away from a solution today as it was more than twenty years ago, when it was first publicly proposed at the beginning of the 1980s.

Biomimetic energy generation so far rather marginal topic in biomimetics

Understanding of biological energy generation processes still insufficient: example photosynthesis

Path dependencies in energy generation sector particularly strong

More biomimetics-specific *obstacles to innovation* could be identified in the case studies:

Biomimetics-specific innovation barriers

a) Complexity match or mismatch

Differences exist, for example, with respect to the complexity of the processes and structures that serve as models. The mechanisms, for example, by which ants – in the quickest and most flexible way – collectively determine the shortest path between their nest and a food source are relatively simple: volatile pheromone tracks are laid down, and then while following these tracks, some very simple rules are followed. The

capability of coping with greater complexities (higher-level) is therefore based on quite simple principles (lower-level). It was possible to transfer these rather less complex mechanisms relatively easily into respective optimisation methods (ant colony optimisation algorithms).
In the case of photosynthesis, however, which serves as a model for biomimetic methods to obtain energy, we are dealing with highly complex structures and processes already at the physical-chemical level; structures that derive energy from photons, pass it on, and then, using a profusion of intermediate reactions, convert it into chemical energy before storing it. The emulation of the photosynthetic process is therefore a far more difficult proposition. The bottom line with respect to driving forces and, in this case in particular, obstacles for biomimetics innovation is clearly the complexity of the model – or more precisely: the degree of complexity in the model massively influences its transferability into technological processes.

b) Degree to which partial functionalities can be isolated and extracted

A further aspect that influences the difficulties of conveying the natural model into the technical application lies in the ability to extract, isolate, and transfer only some aspects of the biological system under consideration. For example, in his attempts at flight based on the model of the flight of birds, Otto Lilienthal's breakthrough only became possible after he had analytically separated the functions of lift and propulsion (thus departing from the imitation of the flapping wing) and concentrated fully on the problem of lift (gliding). In the efforts to develop an artificial photosynthesis, such partial solutions are currently not in sight. It is still not clear whether the process only functions (and should only be labeled as "biomimetic") when the energy is thereby transformed from solar photons into chemically bound energy – ideally with the application of carbon dioxide and water. The structures, materials, and processes of photosynthesis appear to form a single functional unit. If this unity is disrupted by the extraction of individual elements, the overall function appears to be lost. If one thus places biomimetics innovations on a scale with "a copy of nature" at one end and "inspired by nature" at the other, one can conclude that the more tightly one is locked into the integral or functional unity of the natural model, the more difficult the research and development of the process of turning biomimetic ideas into market-ready products and processes become.
This transfer is (fortunately) not always an all-or-nothing situation; but unfortunately the transfer of the desired properties and functions of the biological model is more or less invariably accompanied by taking over of at least some undesirable properties. The best known example

is biodegradable materials intended to remain stable for the duration of their technical utilisation. But also dye-based and other organic solar cells, for example, offer a series of advantages over conventional solar cells thanks to the biological model of photosynthesis they are based on. They can utilize diffuse light, require no scarce or material and energy intensive semiconductor materials, they are light-weight and formable; as a rule, bio-organic materials are also biodegradable. Therefore it is not surprising that the application of organic dyes and other materials in organic solar cells is accompanied by a shorter life cycle and a relatively rapid flattening out of the performance curve over time. While plants and photosynthetic bacteria are able to continually renew the structures that are involved in photosynthesis and thus are able to ensure a constant output over a quite long period of time, organic solar cells lack such possibilities for self-repair and regrowth. As with all innovations, before a biomimetic approach can push aside other non-biomimetic solutions, it must prove to be substantially superior in the sum of its relevant properties and functions. A non-isolatable negative property can thus block the success of highly interesting innovations.

Driving forces for biomimetic innovation

Biomimetics-specific innovation drivers

Biomimetics as an expansion of the field of inquiry

In areas in which there are apparent obstacles to generating solutions on the basis of existing approaches or technologies, a focus on the biomimetic approach seems to hold some value, inasmuch as it expands the problem-solving realm (analogy formation).

Biomimetics – a positive image of new technologies

Biomimetics enjoys an extremely positive public image. The numerous illustrated books and (popular) science books that have appeared in the last few years, as well as a series of television and radio programs on biomimetics, clearly demonstrate the positive image that biomimetics carries. Due to our fascination with the remarkable feats of natural organisms and their structures and processes, biomimetics is viewed as a modern, high-tech matter. At the same time, biomimetics has essentially no negative associations connected to it, because it is nature, after all – which we generally view positively – that serves as the biomimetics model.

Self-x functionalities and bottom-up production paradigm

The fundamental driving forces – which could also and above all be a push forward for biomimetics in the broad sense (learning from nature) and the third strand of nanobionics and nanobiomimetics – may essentially resemble those that have determined the development dynamic of the material sciences and nanotechnologies, as well as hardware and software in the information and communications technologies (self-x) for years. The basic idea of the bottom-up approach – the option of self-organisation or self-growth – opens up interesting new prospects for nanotechnology and biomimetics alike; however, the expected push, the expected dynamic, still tends to be technology driven. The demand side, however, may go on to play an ever greater role in a series of fields of application where the borders of conventional technological approaches increasingly become stumbling blocks, thus intensifying the search for new technological options.

The analysis of the technology forecast studies, the case studies, the discussions in the workshops we conducted, and the interviews make clear that substantial aspects of the technological direction of development can be closely linked to some of the biomimetic guiding principles. In a vast number of areas (robotics, the material sciences, information and communication technologies), biomimetic or bionic aspects play an important role. Biomimetics shows itself to be highly compatible with respect to the development of new technologies, technological developments, and market demands in various science and research areas. The increased demand for adaptiveness of technical systems and the capability to cope with complexity have, at least in part, paved the way for the biomimetic approach. The processes of enquiry in science disciplines such as chemistry, the material sciences, nanotechnology, and information and communications technologies lead in the broadest sense to biomimetic approaches, including multifunctionality and self-healing, among others. In this sense, it is likely that biomimetic approaches will have an increasing influence on the design of the technical systems of the future.

Along with the already mentioned driving forces and obstacles that can be identified along the path of the innovation process, there is a series of other factors that affect the field of biomimetics as a whole:

Biomimetics as scientific field

- *Biomimetic work is (still) characterized by and carried out by a small community and individual persons.*
- *Biomimetics as a field has at its disposal only limited institutional resources. This is true, above all, for the expanded approach of learning from nature.*
- *Companies are currently only able to pursue and implement research projects in the field of biomimetics to a limited extent.*

- *Discipline-based traditions and approaches make the successful implementation of biomimetic approaches more difficult. Biomimetics represents a linking of areas of science that, so far, have been widely separate from one another and still are today; (public) research funding, too, is – at least in the fundamental science areas – often strongly discipline-oriented.*
- *Fascination with biomimetics at the idea level – skepticism with respect to being really scientific and with respect to implementation – that which captures the fascination and enthusiasm of the public at times evokes skepticism among researchers and businesses. Even known exponents of biomimetics still prefer to place themselves – at academic conferences, for example – in their original disciplines.*

Perspectives

There are some good reasons to suggest that biomimetic research and development will play an increasingly important role in the (near) future. On the demand side these are the necessity of adequately dealing with increasing complexity in modern societies and – not least with regard to sustainable development – the necessity of embedding our industrial systems within the carrying capacities of natural systems, that is for more sustainable technologies, production and consumption processes and infrastructures. On the side of technological potentials these are increasing technological possibilities to learn not only from the form and function but also from the control and production processes of natural systems. In the last ten years there were important 'breakthroughs' in fields more led by a broader understanding of 'learning from nature', especially in nanotechnologies, biotechnologies, information and communication technologies and artificial intelligence that opened up new possibilities to 'mimic nature'. Along this line biomimetics may become an interesting experimental learning field for 'converging technologies', possibly the next 'technological revolution'.

7 REFERENCES

7 REFERENCES

ABELE, TH.; KAISER, U.; DRATHEN, H.; WESTERKAMP, D. (2005): *Roadmap Prozess-Sensorik*. In: VDI/VDE-Gesellschaft Mess- und Automatisierungstechnik [Ed.: Verein Deutscher Ingenieure (VDI)] (2005): GMA-Kongress 2005 – Automation als interdisziplinäre Herausforderung [VDI-Berichte Nr. 1883]. Tagung, Baden-Baden, 7.-8.6.2005. VDI Verlag GmbH, Düsseldorf. pp. 695ff.

ABICHT, L.; SCHLICHT, E.; SCHUMANN, U. (2005): *Trendqualifikationen in der Nanotechnologie*. II Eingehende Darstellung. Schlussbericht im Rahmen des Projekts „Ermittlung von Trendqualifikationen im Bereich der Nanotechnologie". Halle.

ANDERSON, D. G.; BURDICK, J. A.; LANGER, R. (2004): *Smart Biomaterials*. Science, Vol. 305, pp. 1923ff.

ANDREESCU, S.; OMOWUNMI, A. S. (2004): *Trends and challenges in biochemical sensors for clinical and environmental monitoring*. In: PURE APPL CHEM, Vol. 76(4), pp. 861-878.

ANTÓN, P. S.; SILBERGLITT, R.; SCHNEIDER, J. (2001): *The Global Technology Revolution*. Bio/Nano/Materials Trends and their Synergies with Information Technology by 2015. RAND Corporation, Santa Monica.

ARGYROS, A. A.; TSAKIRIS, D. P.; GROYER, C. (2004): *Biomimetic centering behavior – Mobile robots with panoramic sensors*. In: IEEE ROBOT AUTOMAT, Vol. 11(4), pp. 21-31.

AYERS, J. (2004): *Underwater walking*. In: ARTHROPOD STRUCT DEV, Vol. 33(3), pp. 347-360.

BALL, P. (2001): *Life's lessons in design*. Nature, Vol. 409, pp. 413-416.

BALL, P. (2005): *Synthetic biology for nanotechnology*. Nanotechnology, Vol. 16, R1-R8.

BANDYOPADHYAY, P. R. (2005): *Trends in biorobotic autonomous undersea vehicles*. In: IEEE Journal of Oceanic Engineering, Vol. 30, Is. 1, pp. 109-139.

BANNASCH, R.; YAKOVLEV, S. (2006): *Maritime Technik: Von der Delphin- Kommunikation zum Tsunami-Frühwarnsystem*. In: Forschungsgemeinschaft Bionik-Kompetenz-Netz e. V. BIOKON (2006): Industriekongress BIONIK 2006 – Innovationsmotor Natur. Berlin. pp. 199ff

BAR-COHEN Y., (ED.) (2005): *Biomimetics: Mimicking and Being Inspired by Biology*. CRC Press.

BAR-COHEN, Y. (2001): *Nanotechnology Using Electroactive Polymers as Artificial Muscles*. MSTnews, International Newsletter on Microsystems and MEMS, Vol. 3/01, pp. 45-46.

BAR-COHEN, Y. (2006): *Introduction to Biomimetics: The Wealth of Inventions in Nature as an Inspiration for Human Innovation*. In: Bar-Cohen, Y. (Ed.) (2006): Biomimetics: Biologically Inspired Technologies. Taylor & Francis, Boca Raton, FL (USA).

BAR-COHEN, Y.; BREAZEAL, C. (ED.) (2003): *Biologically Inspired Intelligent Robots*. SPIE Press, Bellingham.

BENYUS, J. M. (2002): *Biomimicry – innovation inspired by nature*. Harper Perennial. New York (first edition: 1997).

BENSAUDE-VINCENT, B. (2001): *Materials science and engineering: an artificial discipline about to explode?* Download from: http://www.sfc.fr/Material/hrst.mit.edu/hrs/materials/public/Bernadettespaperforhrst_files/intro(bbvhistMSE).htm

BERNHARD, F. (2004): *Technische Temperaturmessung – Physikalische und messtechnische Grundlagen, Sensoren und Messverfahren, Messfehler und Kalibrierung.* Springer Verlag, Berlin et al.

BICO, J.; MARZOLIN, C.; QUÉRÉ, D. (1999): *Pearl drops.* Europhysics Letters 47, pp. 220-226.

BIONA-REPORT 8 (1992): *Technische Biologie und Bionik 1* / 1. Bionik-Kongress, Wiesbaden 1992. Werner Nachtigall (Ed.). Akademie der Wissenschaften und der Literatur, Mainz; G. Fischer, Stuttgart et al.

BLECKMANN, H.; SCHMITZ, H.; VON DER EMDE, G. (2004): *Nature as a model for technical sensors.* In: Journal of Comparative Physiology A, Vol. 190(12), pp. 971-981.

BLÜCHEL, K G. (2006): *Bionik: Wie wir die geheimen Baupläne der Natur nutzen können.* Goldmann, München.

BLÜCHEL, K. G.; MALIK, F. (HRSG.) (2006): *Faszination Bionik: Die Intelligenz der Schöpfung.* Bionik Media, München.

BUNDESMINISTERIUM FÜR BILDUNG UND FORSCHUNG (ED.) (2005): *Regenerative Medizin und Biologie.* Die Heilungsprozesse unseres Körpers verstehen und nutzen, Berlin.

BUNDESMINISTERIUM FÜR BILDUNG UND FORSCHUNG (ED.) (2006A): *Nanotechnologie – Innovationen für die Welt von morgen*, Berlin.

BUNDESMINISTERIUM FÜR BILDUNG UND FORSCHUNG (ED.) (2006B): *Die Hightech-Strategie für Deutschland.* Bonn, Berlin.

Bundesministerium für Bildung und Forschung (Ed.) (2005): Innovationen aus der Natur – Förderkonzept Bionik, Berlin.

BOGATYREV, N. R. (2004): *A "Living" Machine. Journal of Bionics Engineering.* Vol. 1(2), pp. 79-87.

BRACZYK, H. J.; COOKE P.; HEIDENREICH M.(EDS.) (1998): *Regional Innovation Systems.* London

BRESNAHAN, T. F.; TRAJTENBERG M. (1996): *General purpose technologies: 'engines of growth'?,* Journal of Econometrics, Annals of Econometrics 65: pp. 83-108.

BRETTHAUER, G. (2005): *Automatisierungstechnik – Quo vadis? Neun Thesen zur zukünftigen Entwicklung.* Automatisierungstechnik 53 (4-5), pp. 155-157.

BRUCK, H.A.; GERSHON, A.L.; GUPTA, S.K. (2004): *Enhancement of Mechanical Engineering Curriculum to Introduce Manufacturing Techniques and Principles for Bio-inspired Product Development.* Proceedings of IMECE 2004: 2004 ASME International Mechanical Engineering Congress and RD&D Expo, November 2004, Anaheim, Canada.

BSI [BUNDESAMT FÜR SICHERHEIT IN DER INFORMATIONSTECHNIK] (2003): *Kommunikations- und Informationstechnik 2010+3: Neue Trends und Entwicklungen in Technologien, Anwendungen und Sicherheit,* Download from: http://www.bsi.de/literat/studien/trend2010/

CARMENA, J. M.; HALLAM, J. C. T. (2004A): *Narrowband target tracking using a biomimetic sonarhead.* In: ROBOT AUTON SYSTEMS, Vol. 46(4), pp. 247-259.

CARMENA, J. M.; HALLAM, J. C. T. (2004B): *The use of Doppler in Sonarbased mobile robot navigation: inspirations from biology.* In: INFORM SCIENCES, Vol. 161(1-2), pp. 71-94.

CERMAN, Z.; BARTHLOTT, W.; NIEDER, J. (2005): *Erfindungen der Natur: Bionik – Was wir von Pflanzen und Tieren lernen können.* Rowohlt Taschenbuch, Reinbek bei Hamburg.

Dargahi, J.; Najarian, S. (2005): *Advances in tactile sensors design / manufacturing and its impact on robotics applications – a review.* In: Industrial Robot: An International Journal, Vol. 32, Is. 3, pp. 268-281.

Dario, P.; Carrozza, M. C.; Guglielmelli, E.; Laschi, C.; Menciassi, A.; Micera, S.; Vecchi, F. (2005): *Robotics as a future and emerging technology: Biomimetics, cybernetics, and neurorobotics in European projects.* In: IEEE ROBOT AUTOM MAG, Vol. 12 (2), pp. 29-45.

David, P. A. (1987): *Some new standards for the economics of standardization in the information age,* in: Dasgupta, Partha., Stoneman, Paul. (ed.): Economic policy and technological performance, pp. 206-239.

de Casto, L. N. (2005): *Biological Inspired Computing.* Presentation for the International Symposium on Bio-Inspired Computing. Johor, MY, 6. September 2005.

Deville, S; Saiz, E.; Nalla, R. K.; Tomsia, A. P. (2006): *Freezing as a Path to Build Complex Composites*, SCIENCE, Vol. 311, pp. 515-518.

Dickinson, M. H. (1999): *Bionics - Biological insight into mechanical design.* Proceedings of the National Academy of Sciences USA 96(25), pp. 14208-14209.

Dolata, U. (2004): *Unfassbare Technologien, internationale Innovationsverläufe und ausdifferenzierte Politikregime.* Perspektiven nationaler Technologie- und Innovationspolitiken. artec-paper Nr. 110 März 2004.

Dosi, G. (1982): *Technological Paradigms and Technological Trajectories: A Suggested Interpretation of the Determinants and Directions of Technical Change*, Research Policy, Vol. 11, Nr. 3, pp. 47-162.

Edquist, C. (2001): *Systems of innovation*, in Michie (ed.), A Reader's Guide to the Social Sciences, Fitzroy Dearborn/Routledge, London.

Engin, M.; Demirel, A.; Engin, E. Z.; Fedakar, M. (2005): *Recent developments and trends in biomedical sensors.* In: MEASUREMENT, Vol. 37(2), pp. 173-188

Enterprise Management Associates (2006): *Practical Autonomic Computing.* Roadmap to Self Managing Technology. A White Paper prepared for IBM. Download from: www-03.ibm.com/autonomic/pdfs/AC_Practical-RoadmapWhitepaper_051906.pdf

ESA [European Space Agency] (Ed.) (2004): *Biomimicry – A Review* (prepared by Mark Ayre), Download from: http://www.esa.int/gsp/ACT/doc/ACT_ RPR_6000_BIO_Review_AYRE.pdf

Filliat, D.; Girard, B.; Guillot, A.; Khamassi, M.; Lachèze, L.; Meyer, J.-A. (2004): *State of the Artificial Rat Psikharpax.* In: Schaal, S. et al. (eds.) (2004): From Animals to Animats 8 – Proceedings of the Eighth International Conference on Simulation of Adaptive Behavior. MIT Press, Cambridge et al., pp. 3-12.

Foresight (Ed.) (2000): *Materials: Shaping our Society – Materials Panel consultation recommendations.* Download from: http://www.foresight.gov. uk/Previous_Rounds/Foresight_1999__2002/Materials/Reports/Materials%20Shaping%20our%20Society/Report.pdf

Forth, E.; Schewitzer, E. (1976): *Bionik.* [Meyers Taschenlexikon]. VEB Bibliographisches Institut, Leipzig.

Franceschini, N. (2004): *Visual guidance based on optic flow: a biorobotic approach.* In: Journal of Physiology-Paris, Vol. 98(1-3), pp. 281-292.

FRATZL, P.; GUPTA, H.S.; PASCHALIS, E.P.; ROSCHGER, P. (2004): *Structure and mechanical quality of the collagen-mineral nano-composite in bone*, J.Mater.Chem., 14, pp. 2115-2123.

FREEMAN, C. (1987): *The 'National System of Innovation' in Historical Perspective.* In: D. Archibugi & J. Michie (Eds): Technology, Globalisation, and Economic Performance, Cambridge University Press, Cambridge & New York.

FRITZ, M.; GRATHWOHL, G.; VON GLEICH, A.; KUNTZ, M.; ZINSMEISTER, K. (2005): *Perlmutt – Vorbild für nachhaltig zukunftsfähige Werkstoffe* (www.tecdesign.uni-bremen.de/FG10/documente/PerlmuttBericht.doc

FROMM, J. (2005): *Ten questions about emergence.* Technical report. Online: http://arxiv.org/abs/nlin.AO/0509049.

FULK, K. (2002): *Biological Computing.* ISRC Future Technology Topic Brief, November 2002. Download from: http://www.uhisrc.com/futuretech.htm

FURUKAWA, T. (2004): *Biological Imaging and Sensing.* Springer Verlag, Berlin et al.

GAFFRON, S. (2006): *Grüne Luftschlösser.* In: Technology Review [Deutsche Ausgabe], Vol. 2006, Is. 11, pp. 72-73.

GAUSEMEIER, J.; FRANK, U.; SCHULZ, B. (2005): *Domänenübergreifende Spezifikation der Prinziplösung selbstoptimierender Systeme unter Berücksichtigung der auf das System wirkenden Einflüsse.* In: VDI/VDE-Gesellschaft Mess- und Automatisierungstechnik [Ed.: VDI Wissensforum IWB GmbH] (2005): Mechatronik 2005 – Innovative Produktentwicklung [VDI-Berichte Nr. 1892.1/2]. Tagung, Wiesloch, 1.-2.6.2005. VDI Verlag GmbH, Düsseldorf. pp. 315ff.

GÉRARDIN, L. (1972): *Natur als Vorbild: Die Entdeckung der Bionik.* [Bücher des Wissens]. Fischer Taschenbuch, Frankfurt a. M.

GLEICH, A. VON (1989): *Der wissenschaftliche Umgang mit der Natur – Über die Vielfalt harter und sanfter Naturwissenschaften* (Campus Forschung), Frankfurt / New York.

GLEICH, A. VON (1998): *Was können und sollen wir von der Natur lernen?* In: von Gleich, A. (Ed.): Bionik – Ökologische Technik nach dem Vorbild der Natur? Teubner Stuttgart (second edition 2001).

GLEICH, A. VON (1998/99): *Ökologische Kriterien der Technik- und Stoffbewertung: Integration des Vorsorgeprinzips.* Umweltwissenschaften und Schadstoff-Forschung – Zeitschrift für Umweltchemie und Ökotoxikologie, Vol. 10(6), 11(1/2).

GLENN, J.C.; GORDON, T.J. (2006): *2006 State of the Future.* American Council for the United Nations.

GOODSELL, D.S. (2004): *Bionanotechnology: Lessons from Nature.* Hoboken (New Jersey).

GORBUNOV, V.; FUCHIGAMI, N.; STONE, M.; GRACE, M.; TSUKRUK, V. V. (2002): *Biological thermal detection: Micromechanical and microthermal properties of biological infrared receptors.* In: BIOMACROMOLECULES, Vol. 3(1), pp. 106-115.

GRATHWOHL, G. (2004): *Neue Keramik: Aufbruch in Biosphäre und Nanowelt*, Bremer Universitätsgespräche, Dokumentation des 16. BUG am 30. und 31. Okt. 2003, Universitätsverlag Aschenbeck & Isensee, Bremen.

GREGUSS, F. (1988): *Patente der Natur: Technische Systeme in der Tierwelt – Biologische Systeme als Modelle für die Technik.* Quelle & Meyer, Heidelberg/ Wiesbaden.

GRÜNDLER, P. (2004): *Chemische Sensoren – Eine Einführung für Naturwissenschaftler und Ingenieure.* Springer Verlag, Berlin et al.

GRUPP, H.; DOMINGUEZ-LACASA I.; FRIEDRICH-NISHIO M. (2002): *Das deutsche Innovationssystem seit der Reichsgründung.* Indikatoren einer nationalen Wissenschafts- und Technikgeschichte in unterschiedlichen Regierungs und Gebietsstrukturen (= Technik, Wirtschaft und Politik. Schriftenreihe des Fraunhofer-Instituts für Systemtechnik und Innovationsforschung ISI; 48), Heidelberg: Physica-Verlag.

HAMPP, N. (2005): *Bio – Nano – NanoBio: Chancen, Risiken, Visionen.* Folienvortrag auf dem 2. Nanotechnologieforum Hessen am 16. November 2005 in Hanau.

HAMPP, N.; NOLL F. (2003): *Nanobionics – from Molecules to Applications.* 293. WE Heraeus Seminar. Physik Journal 2(2), p. 56.

HERR, H.; DENNIS, R.G. (2004): *A Swimming Robot Actuated by Living Muscle Tissue.* Journal of NeuroEngineering and Rehabilitation, Vol. 1(6).

HERR, H; WHITELEY G.; CHILDRESS D. (2003): *Cyborg Technology – Biomimetic Orthotic and Prosthetic Technology.* In: Bar-Cohen, Y.; C. Breazeal (Eds.): Biologically Inspired Intelligent Robots. SPIE Press, Bellingham, Washington.

HEYNERT, H. (1976): *Grundlagen der Bionik.* Dr. Alfred Hüthig, Heidelberg.

HIGGINS, C. M.; AND PANT, V. (2004): *A biomimetic VLSI sensor for visual tracking of small moving targets.* In: IEEE TRANS CIRCUIT SYST-I, Vol. 51(12), pp. 2384-2394.

HILL, B. (ED.) (2006): *Bionik: Lernen von der Natur.* [Natur – Mensch – Technik]. Duden Paetec. Berlin/Frankfurt a. M.

HONSEL, G. (2006): *Wettkampf der Profile.* In: Technology Review [Deutsche Ausgabe], Vol 2006, Is. 11, pp. 74-75.

HUMPL, S.; NEUBERGER, R.; NINDL, S.; ÖLLER, M.M.; SCHAMSCHULE, R. (2006): *Aus- und Weiterbildung in Nanowissenschaften und Nanotechnologien* –Bestands- und Bedarfserhebung in Österreich. Wien.

ICHBIAH, D. (2005): *Roboter: Geschichte, Technik, Entwicklung.* Knesebeck Verlag, München.

IEA [INTERNATIONAL ENERGY AGENCY] (2006): *Energy technology perspectives: in support of the G8 plan of action*; scenarios & strategies to 2050. Paris: OECD.

IEEE [INSTITUTE OF ELECTRICAL AND ELECTRONICS ENGINEERS] (2004): *IEEE International Conference on Robotics and Biomimetics 2004.* Piscataway, N.J. Download from: http:// ieeexplore.ieee.org/servlet/opac?punumber=10204

IFMO [INSTITUT FÜR MOBILITÄTSFORSCHUNG] (2006): *Zukunft der Mobilität – Szenarien für das Jahr 2025.* Berlin.

INSTITUT FÜR TECHNIKFOLGENABSCHÄTZUNG [ITA] (2006): *Bionik-Potential in Österreich.* Studie im Auftrag des Bundesministeriums für Verkehr, Innovation und Technologie Österreich, Wien.

JAAX, K. N. AND HANNAFORD, B. (2004): *Mechatronic design of an actuated biomimetic length and velocity sensor.* In: IEEE TRANS ROBOTICS AUTOMAT, Vol. 20(3), pp. 390-398.

JACOB K.; BEISE, M.; BLAZEJCZAK, J.; EDLER, D.; HAUM, R.; JÄNICKE, M.; LOEW, T.; PETSCHOW, U.; RENNINGS, K. (2005): *Lead Markets of Environmental Innovations.* Physica-Verlag, Heidelberg.

JACKSON, R. G. (2004): *Novel Sensors and Sensing.* IOP Publishing Ltd., Bristol, Philadelphia.

JOACHIM, C. (2005): *To be nano or not to be nano?* Commentary. Nature materials, Vol. 4, pp. 107-109.

JOHNSON, A., T., SCHREUDERS, P. D. (2003): *Bringing life to engineering: biological engineering at the graduate level.* European Journal of Engineering Education, 28(1), pp. 37-46.

KANOUN, O.; TRÄNKLER, H.-R. (2004): *Sensor Technology Advances and Future Trends.* In: IEEE TRANS INSTRUM MEAS, Vol. 53(6), pp. 1497-1501.

KASINGER, H. (2005): *Ein MDA-basierter Ansatz zur Entwicklung von Organic Computing Systemen.* TR 2005-08 Institut für Informatik Universität Augsburg. April 2005.

KEMP, R. (2001): *Possibilities for a Green Industrial Policy from an Evolutionary Technology Perspective.* In: M. Binder, M. Jänicke, and U. Petschow (Eds.): Green Industrial Restructuring - International Case Studies and Theoretical Interpretations, Springer: Berlin/ Heidelberg/ New York, pp. 151-169.

KENNY, T. (2005): *Basic Sensor Technology.* In: Wilson, J. S. (Ed.) (2005): Sensor Technology Handbook. Newnes (Elsevier), Amsterdam et al. pp. 1-15.

KEPHART, J., D. CHESS. (2003): *The Vision of Autonomic Computing.* Computer 36, 1 (Jan. 2003), pp. 41-50.

KESEL, A. (2006): *Bionik.* Fischer Taschenbuch, Frankfurt a. M.

KEY TECHNOLOGIES EXPERT GROUP (2005): Creative System Disruption. Towards a Research Strategy Beyond Lisbon. Draft Final Report. Download from: http://cordis.europa.eu/foresight/kte_expert_group_2005.htm

KIENER, U. (2005): *Zukunft Engineering.* Eine Expertenbefragung in der Schweiz. Studie im Auftrag des ETH-Rates und des Bundesamtes für Berufsbildung und Technologie. Winterthur.

KÖNIG, A.; EBERHARDT, M.; DÖGE, J.; WENZEL, R.; SKRIBANOWITZ, J. (2001): *Perspektive des Entwurfs verlustleistungsarmer integrierter Signalverarbeitung für intelligente Mikrosensoren in lernender Architektur.* In: VDI/VDE-Gesellschaft Mess- und Automatisierungstechnik [Ed.: Verein Deutscher Ingenieure (VDI)] (2001): Automatisierungstechnik im Spannungsfeld neuer Technologien [VDI-Berichte Nr. 1608]. Tagung, Baden-Baden, 22./23. Mai 2001. VDI Verlag GmbH, Düsseldorf. pp. 61ff

KRAISMER, L. P. (1967): *Bionik: Eine neue Wissenschaft.* [Kleine naturwissenschaftliche Bibliothek, Reihe Physik, Vol. 4]. [second edition]. B. G. Teubner Verlagsgesellschaft, Leipzig.

KRETSCHMER, TH. (2006): *Technologiefelder der Zukunft – Gesamtbild und Megatrends.* Folien zum Vortrag am 4. Innovationstag in Innsbruck, 30. November 2006.

KÜPPERS, U. (2004): *Versuch einer Annäherung: Bionik und Wirtschaftlichkeit.* In: Biologie in unserer Zeit, Vol. 34, Is. 5, pp. 316-323.

KÜPPERS, U.; TRIBUTSCH, H. (2002): *Verpacktes Leben – Verpackte Technik: Bionik der Verpackung.* Wiley VCH, Weinheim.

LANGER, R.; TIRRELL, D. A. (2004): *Designing Materials for Biology and Medicine.* Nature, Vol. 428, pp. 487-492.

LEE, L.P.; SZEMA, R. (2005): *Inspirations from Biological Optics for Advanced Photonic Systems*, SCIENCE, Vol. 310, pp. 1148-1150.

7

LI, W.; FARREL, J. A.; CARDE, R. T. (2001): *Tracking of fluid-advected odor plumes: Strategies inspired by insect orientation to pheromone.* In: ADAPT BEHAV, Vol. 9(3-4), pp. 143-170.

LIU, S. C.; TOMIZUKA, M; ULSOY, A. G. (2005): *Challenges and opportunities in the engineering of intelligent systems.* In: SMART STRUCT SYST, Vol. 1(1), pp. 1-12.

LUNDVALL, B.-A. (ED.) (1992): *National Systems of Innovation: Towards a Theory of Innovation and Interactive Learning*, Pinter, London.

MACIVER, M. A.; NELSON, M. E. (2001): *Towards a biorobotic electrosensory system.* In: AUTON ROBOTS, Vol. 11(3), pp. 263-266.

MACKENSEN, E.; KUNTZ, W. (2004): *Intelligente, autarke Mikrosysteme für drahtlose Sensor-Aktor-Netzwerke.* In: VDI/VDE-Gesellschaft Mess- und Automatisierungstechnik [Hrsg.:Verein Deutscher Ingenieure (VDI)] (2004): Sensoren und Messsysteme 2004 [VDI-Berichte Nr. 1829]. Tagung, Ludwigsburg, 15.-16.3.2004. VDI Verlag GmbH, Düsseldorf. pp. 307ff.

MANN, S. (ED) (1996): *Biomimetic Materials Chemistry,* VCH Weinheim.

MANN, S.; WERBB, J.; WILLIAMS, R. (EDS) (1989): *Biomineralization, Chemical and Biological Perspectives*, VCH Weinheim.

MARGUERRE, H. (1991): *Bionik: Von der Natur lernen.* Siemens Aktiengesellschaft, Berlin/München.

MATTHECK, C. (1999): *Stupsi erklärt den Baum – ein Igel lehrt die Körpersprache der Bäume.* Verlag Forschungszentrum Karlsruhe, Vol. 3.

MATTHECK, C. (2003): *Warum alles kaputt geht – Form und Versagen in Natur und Technik.* Verlag Forschungszentrum Karlsruhe.

MATTHECK, C. (2006): *Verborgene Gestaltgesetze der Natur.* Verlag Forschungszentrum Karlsruhe.

MAYER, G. (2005): *Rigid Biological Systems as Models for Synthetic Composites,* SCIENCE, Vol. 310, pp. 1144-1147.

MCGRUER, N. E.; ADAMS, G. G.; TRUONG, T. Q.; BARNES, T. G.; LU, X; ACEROS, J. C. (2002): *Biomimetic Flow and Contact/Bending MEMS Sensors.* In: Ayers, J. et al. (Eds.) (2002): Neurotechnology for Biomimetic Robots. MIT Press, Cambridge (Massachusetts). pp. 13ff.

MEFI [MINISTÈRE DE L'ECONOMIE, DES FINANCES ET DE L'INDUSTRIE] (ED.) (2006): *Technologies clés 2010.* Download from : http://www.tc-2010.fr/

MEIER, H.; SMAJIC, H. (2005): *Entwurfs- und Projektierungswerkzeuge für intelligente und verteilte Aktor-/Sensorsysteme.* In: VDI/VDE-Gesellschaft Mess- und Automatisierungstechnik [Hrsg.: VDI Wissensforum IWB GmbH] (2005): Mechatronik 2005 – Innovative Produktentwicklung [VDI-Berichte Nr. 1892.1/2]. Tagung, Wiesloch, 1.-2.6.2005. VDI Verlag GmbH, Düsseldorf. pp. 1049ff.

MEYER, J.-A.; GUILLOT, A.; GIRARD, B.; KHAMASSI, M.; PIRIM, P.; BERTHOZ, A. (2005): *The Psikharpax project: towards building an artificial rat.* In: ROBOT AUTON SYST, Vol. 50 (4), pp. 211-223.

MICROSOFT (2005): *Towards 2020 Science Report.* Download from: http://research.microsoft.com/towards2020science/downloads/T2020S_ReportA4.pdf

MINAI, A. A.; BRAHA, D.; BAR-YAM, Y. (2006): *Complex Engineered Systems: A New Paradigm.* In: Braha, D., Minai, A.A., Bar-Yam, Y., Complex Engineered Systems, Cambridge, Massachusetts: Springer, pp. 1-21.

MÜLLER-SCHLOER, C. (2004): *Organic computing – on the feasibility of controlled emergence.* In: IEEE/ACM/IFIP International Conference on Hardware/Software Codesign and System Synthesis (CODES+ISSS), Stockholm. pp. 2-5.

MÜLLER-SCHLOER, C.; VON DER MALSBURG, C.; WÜRTZ, R. P. (2004): *Organic Computing.* Informatik Spektrum 4 (2004), pp. 2-6.

NACHTIGALL, W. (1998): *Bionik: Grundlagen und Beispiele für Ingenieure und Naturwissenschaftler.* Springer, Berlin et al.

NACHTIGALL, W.; BLÜCHEL, K. G. (2003): *Das große Buch der Bionik: Neue Technologien nach dem Vorbild der Natur.* [Sonderausgabe]. Deutsche Verlags-Anstalt, München/Stuttgart.

NACHTIGALL, WERNER (2002): *Bionik. Grundlagen und Beispiele für Ingenieure und Naturwissenschaftler.* Springer, Berlin. Vol. 2.

NATIONAL RESEARCH COUNCIL OF CANADA (ED.), (2003): *Technology Foresight Pilot Project.* BioSystemics Knowledge Manager's Synthesis Project. Download from: www.moyak.com/researcher/nrc/BioSystemics.pdf

NELSON, R. (ED.) (1993): *National Innovation Systems: A Comparative Analysis.* New York, Oxford.

NIDCR [NATIONAL INSTITUTE OF DENTAL AND CRANIOFACIAL RESEARCH] (ED.) (2005): *Biomimetics and Tissue Engineering.* Download from: www.nidcr.nih.gov

NIMS [NATIONAL INSTITUTE FOR MATERIALS SCIENCE] (ED.), (2005): *Materials Science Outlook 2005.* Download from: http://www.nims.go.jp/eng/news/outlook/index.html

NISTEP [NATIONAL INSTITUTE OF SCIENCE AND TECHNOLOGY POLICY] (ED.) (2005): *The 8th Science and Technology Foresight Survey.* Delphi Analysis. NISTEP Report No. 97, Download from: http://www.nistep.go.jp/achiev/ftx/eng/rep097e/idx097e.html

NORDMANN, A. (ED.) (2004): *Converging Technologies – Shaping the Future of European Societies.* Report from the High Level Expert Group on "Foresighting the New Technology Wave". Download from: www.ntnu.no/2020/final_report_en.pdf

NSANZE, F. (2005): *ICT Implants in the Human Body. A Review.* The European Group on Ethics in Science and New Technologies to the European Commission. Brussels.

OECD [ORGANISATION FOR ECONOMIC CO-OPERATION AND DEVELOPMENT] (ED.) (2006): *Infrastructure to 2030.* Telecom, Land Transport, Water and Electricity. Paris: OECD.

OERTEL, D.; GRUNWALD, A. (2006): *Potenziale und Anwendungsperspektiven der Bionik.* Vorstudie des Büro für Technikfolgenabschätzung beim Deutschen Bundestag (TAB). Arbeitsbericht 108. Download from: http:// www.tab.fzk.de/de/projekt/zusammenfassung/ab108.pdf

PAUN, GH. (2004): *Bio-Inspired Computing Paradigms (natural computing).* Download from: http://upp.lami.univ-evry.fr/Documents/Paun/paun-biocom.pdf

PETSCHOW, U.; VON GLEICH, A.; HAUM, R.; LEVINA, O. (2005): *Potenziale und Anwendungsperspektiven der Bionik.* Themenfeld 2: Die Nähe zur Natur als Chance und als Risiko. Studie im Auftrag des Deutschen Bundestages, vorgelegt dem Büro für Technikfolgenabschätzung beim Deutschen Bundestag, Berlin/Bremen 15/11/2005.

PETTAZZI, L.; AYRE, M.; IZZO, D. (2005): *Self-Assembly in Space Using Behaviour-Based Intelligent Components.* Internal Report. Prepared for ESA.

POPPER, S. W.,; WAGNER, C. S.,; LARSON, E. V. (1998): *New Forces At Work: Industry Views Critical Technologies.* RAND Corporation, Santa Monica.

PRIAMI, C. (2005): *Complexity and Systemics.* Expert Report for the Expert Group "Key Technologies for Europe". Download from: http://cordis.europa.eu/foresight/kte_expert_group_2005.htm

PROCK, J. (2004): *Selbstüberwachung und übergreifende Überwachung von Sensorsystemen der Verfahrenstechnik.* In: VDI/VDE-Gesellschaft Mess- und Automatisierungstechnik [Hrsg.: Verein Deutscher Ingenieure (VDI)] (2004): Sensoren und Messsysteme 2004 [VDI-Berichte Nr. 1829]. Tagung, Ludwigsburg, 15.-16.3.2004. VDI Verlag GmbH, Düsseldorf. pp. 179ff.

PROKOP, A. (2001): *Bioartificial Organs in the 21st Century: Nanobiological Devices.* Annals of the New York Academy of Sciences, Vol. 944.

PUCCINELLI, D.; HAENGGI, M. (2005): *Wireless Sensor Networks: Applications and Challenges of Ubiquitous Sensing.* In: IEEE Circuits and Systems Magazin, Vol. 5(3), pp. 19-29.

RGTF [RESEARCH GROUP OF TECHNOLOGY FORESIGHT] (ED.) (2005): *China's Report of Technology Foresight.* Summary. National Research Center for Science and Technology for Development, People's Republic of China. Download from: http://www.foresight.org.cn/cbw/national%20technology%20foresight_03_04-051101.pdf

RIECHEY, JULIANE (2008): *Das implizite Versprechen der Bionik – Untersuchung auf Evidenz, Verbreitung, sowie Methoden zur Bestimmung der Validität,* Diplomarbeit Universität Bremen.

ROCO, M. C.; BAINBRIDGE, W. S. (EDS.), (2002): *Converging Technologies for Improving Human Performance.* Nanotechnology, Biotechnology, Information Technology and Cognitive Science. National Science Foundation. Download from: http://www.wtec.org/ConvergingTechnologies/

RODRIGUEZ-MOZAZ, S.; LOPEZ DE ALDA, M. J.; MARCO, M.-P.; BARCELO, D. (2005): *Biosensors for environmental monitoring – A global perspective.* In: TALANTA, Vol. 65(2), pp. 291-297.

ROSSMANN, T.; TROPEA, C. (2005): *Vorwort der Herausgeber.* In: Rossmann, T.; Tropea, C. (Hrsg.) (2005): Bionik: Aktuelle Forschungsergebnisse in Natur-, Ingenieur- und Geisteswissenschaft. Springer, Berlin et al.

RÖTZER, F. (2001): *Robotfisch mit Froschmuskeln.* Telepolis 22.02.2001. Download from: www.heise.de/tp/

RUFFIER, F.; FRANCESCHINI, N. (2005): *Optic flow regulation: the key to aircraft automatic guidance.* In: ROBOT AUTON SYSTEMS, Vol. 50(4), pp. 177-194.

RÜHLE, M.; DOSCH, H.; MITTEMEIJER, E. J.; VAN DE VOORDE, M. H. (EDS.) (2001): *European White Book on Fundamental Research in Materials Science.* Max-Planck-Institut für Metallforschung Stuttgart. Download from: http://www.mpg.de/english/illustrationsDocumentation/documentation/europWhiteBook/index.html

SALO, A.; CUHLS, K. (2003): *Technology Foresight – Past and Future.* Journal of Forecasting, Vol. 22, pp. 79-82.

SANCHEZ, C. (COORD.) (2001): *Biomimétisme et Matériaux.* Arago 25. OFTA Paris.

SANCHEZ, C.; ARRIBART, H.; GIRAUD, G.; MARIE, M. (2005): *Biomimetism and Bioinspiration as Tools for the Design of Innovative Materials and Systems.* Nature Materials, Vol. 4, pp. 277-288.

SARTORIUS, C.; ZUNDEL, S. (EDS.) (2005): *Time Strategies and Environmental Policy.* Edward Elgar Publishers (Celtenham).

SAXL, O. (2005): *Nanotechnology – A Key Technology for the Future of Europe.* Expert Report for the Expert Group "Key Technologies for Europe". Download from: http://cordis.europa.eu/foresight/kte_expert_group_2005.htm

SCHANZ, G. W. (2003): *Sensortechnik aktuell – Ausgabe 2003: Trends, Produkte und Entscheidungshilfen.* Oldenbourg Industrieverlag, München.

SCHANZ, G. W. (2004): *Sensortechnik aktuell – Ausgabe 2004: Trends, Produkte und Entscheidungshilfen.* Oldenbourg Industrieverlag, München.

SCHANZ, G. W. (2005): *Sensortechnik aktuell – Ausgabe 2005: Trends, Produkte und Entscheidungshilfen.* Oldenbourg Industrieverlag, München.

SCHANZ, G. W. (2006): *Sensortechnik aktuell – Ausgabe 2006: Trends, Produkte und Entscheidungshilfen.* Oldenbourg Industrieverlag, München.

SCHENATO, L.; WU, W. C.; SASTRY, S. S. (2004): *Attitude control for a micromechanical flying insect via sensor output feedback.* In: IEEE TRANS ROBOTICS AUTOMAT, Vol. 20(1), pp. 93-106.

SEEMAN, N. C.; BELCHER, A. M. (2002): *Emulating biology: Building nanostructures from the bottom up.* PNAS, Vol. 99, Suppl. 2, pp. 6451-6455.

SELVERSTON, A. I. (2002): *Advances in the Development of Biomimetic Sensors, Actuators, and Controllers.* In: Ayers, J. et al. (Eds.) (2002): Neurotechnology for Biomimetic Robots. MIT Press, Cambridge et al., pp. 4-10.

SETTLES, G. S. (2005): *Sniffers: Fluid-dynamic sampling for olfactory trace detection in nature and homeland security – the 2004 Freeman Scholar Lecture.* In: J FLUID ENG, Vol. 129(2), pp. 189-218.

SHAHINPOOR, M. (2003): *Ionic polymer-conductor composites as biomimetic sensors, robotic actuators and artificial muscles – an overview.* In: ELECTROCHIM ACTA, Vol. 48 (14-16), pp. 2343-2353.

SHAHINPOOR, M.; KIM, K. J. (2004): *Ionic polymer-metal composites: IV. Industrial and medical applications.* In: SMART MATER STRUCT, Vol. 14(1), pp. 197-214

SHELL INTERNATIONAL (ED.) (2005): *Shell global scenarios to 2025 : the future business environment; trends, trade-offs and choices.* London: Shell International Ltd.

SHIN H.; JO, S.; MIKOS, A. G. (2003): *Biomimetic materials for tissue engineering.* Biomater, Vol. 24, pp. 4353-4364.

SIEMENSFORUM/LANDESMUSEUM FÜR TECHNIK UND ARBEIT (ED.) (1999): *Bionik: Zukunfts-Technik lernt von der Natur* (Eine gemeinsame Wanderausstellung des SiemensForums München/Berlin und des Landesmuseums für Technik und Arbeit in Mannheim). [veränderter Nachdruck, Original von 1998]. SiemensForum, München/Berlin & Landesmuseum für Technik und Arbeit, Mannheim.

SILBERGLITT, R.; ANTÓN, P. S.; HOWELL, D. R.;, WONG, A. (2006): *The Global Technology Revolution 2020, In-Depth Analyses.* Bio/Nano/Materials/Information Trends, Drivers, Barriers, and Social Implications. Rand Corporation, Santa Monica.

SINÁK, P.; VAŠÁK, J.; HIROTA, K. (2004): *Machine Intelligence: Quo Vadis?* World Scientific Publishing, Singapore.

SKORDOS, A.; CHAN P. H.; VINCENT, J. F. V., JERONIMIDIS, G. (2002): *A novel strain sensor based on the companiform sensillum of insects.* In: PHIL TRANS ROY SOC LONDON A, Vol. 360(1791), pp. 239-253.

SÖRENSEN M. H. (2004): *Ambient Ecologies – Toward biomimetic IT.* Ph.D. Dissertation, IT University of Copenhagen.

SÖRENSEN, M. H. (2003): *Ambient Intelligence Ecologies.* Download from: http://www.itu.dk/people/megel/ambientIntelligenceEcologies.pdf

SPECHT, G. (2005): *Technologie- und marktorientierte Entwicklung von Bionik-Produkten.* In: Rossmann, T., Tropea, C. (Eds.) (2005): Bionik: Aktuelle Forschungsergebnisse in Natur, Ingenieur und Geisteswissenschaft. Springer, Berlin et al. pp. 247-266.

SPECK, O. (2006): *Schulversuche zur Bionik.* In: Hochschule Bremen (Ed.), Tagungsunterlagen. Bionik-Kongress in Bremen „Patente aus der Natur", 27/28 Oktober 2006.

STEINFELDT, M.; VON GLEICH A., PETSCHOW U.; HAUM R. (2007): *Nanotechnologies, Hazards and Resource Efficiency : A three-tiered approach to assessing the implications of nanotechnology and influencing its development.* Springer Verlag.

STEVENS, M. M.; GEORGE, J. H. (2005): *Exploring and Engineering the Cell Surface Interface.* Science, Vol. 310.

STIBOR, I. (ED.) (2005): *Anion Sensing.* Springer Verlag, Berlin et al.

SZENTPÉTERY, V. (2006): *Werkzeuge aus der Natur.* In: Technology Review [German edition], Vol. 2006, Is. 11, pp. 62-67.

TNS [TNS INFRATEST WIRTSCHAFTSFORSCHUNG] (2004): *Horizons 2020. A thought-provoking look at the future.* München.

Toko, K. (2000): *Biomimetic sensor technology.* Cambridge University Press, Cambridge et al.

TOKO, K.; HABARA, M. (2005): *Taste sensor.* In: CHEM SENSES, Vol. 30, Suppl. 1, pp. i256-i257

TRÄNKLER, H.-R.; OBERMEIER, E. (HRSG.) (1998): *Sensortechnik – Handbuch für Praxis und Wissenschaft.* Springer Verlag, Berlin et al.

TRUMLER, W. (2006): *Organic Ubiquitous Middleware.* Dissertation at the University of Augsburg.

UMSICHT [FRAUNHOFER-INSTITUT UMWELT-, SICHERHEITS- UND ENERGIETECHNIK] (ED.) (2005): *Bionik als Technologievision der Zukunft.* Studie im Auftrag des Büro für Technikfolgenabschätzung beim Deutschen Bundestag (TAB). Download from: http://dip.bundestag.de/btd/16/037/1603774.pdf

VAN LIESHOUT, M.; ENZING, C.; HOFFKNECHT, A.; HOLTMANSPOTTER, D.; NOYONS, E. (2006): *Converging Applications Enabling the Information Society.* Trends and Prospects of the Convergence of ICT with Cognitive Science, Biotechnology, Nanotechnology and Material Sciences. Download from: http://fiste.jrc.es/pages/horizontal.htm

VEREIN DEUTSCHER INGENIEURE, VDI-TECHNOLOGIEZENTRUM PHYSIKALISCHE TECHNOLOGIEN (ED.) (NEUMANN, D.) (1993): *Technologieanalyse Bionik.* Analyse & Bewertung Zukünftiger Technologien. Düsseldorf.

VEREIN DEUTSCHER INGENIEURE, VDI-TECHNOLOGIEZENTRUM (ED.) (SEILER, P.; HOLTMANNSPÖTTER, D.; ALBERTSHAUSER, U.) (2004): *Internationale Technologieprognosen im Vergleich.* Übersichtsstudie. Zukünftige Technologien, Düsseldorf. VDI-Technologiezentrum (Ed.) (Wevers, M., Wechsler, D.) (2002): Nanobiotechnologie I: Grundlagen und technische Anwendungen molekularer, funktionaler Biosysteme. Zukünftige Technologien Nr. 38. Düsseldorf.

Verein Deutscher Ingenieure, VDI-Technologiezentrum (Ed.) (Holtmannspötter, D.; Rijkers-Defrasne, S.; Glauner, C.; Korte, S.; Zweck, A.) (2006): *Aktuelle Technologieprognosen im internationalen Vergleich.* Übersichtsstudie. Zukünftige Technologien, Düsseldorf.

Versweyveld, L. (2004): *U.S. Department of Veterans Affairs to fund new centre for advanced limb-loss research.* Virtual Medical World, 16.12.2004.

Vincent, J. F. V.; Bogatyreva, O. A.; Bogatyrev, N. R.; Bowyer, A; Pahl, A. (2006): *Biomimetics: its practice and theory.* In: Journal of the Royal Society Interface, Vol. 3, pp. 471-482.

Viollet, S.; Franceschini, N. (2005): *A high speed gaze control system based on the Vestibulo-Ocular Reflex.* In: ROBOT AUTON SYSTEMS, Vol. 50(4), pp. 147-161.

Walter, A. (2005): *Technologietransfer.* In: Albers, S.; Gassmann, O. (Ed.) (2005): Handbuch Technologie- und Innovationsmanagement. Gabler, Wiesbaden. pp. 101-118.

Webb, B.; Harrison, R. R.; Willis, M. A. (2004): *Sensorimotor control of navigation in arthropod and artificial systems.* In: ARTHROPOD STRUCT DEV, Vol. 33(3), pp. 301-329.

Weiss, P. (2003): *Channelling light in the deep sea.* Science news (164).

Wen, X.; Shi, D.; Zhang, N. (2005): *Applications of Nanotechnology in Tissue Engineering.* In: Nalwa, H. S. (Ed.): Handbook of Nanostructured Biomaterials and Their Applications in Nanobiotechnology, Vol. 1, pp. 1-23.

Wengenroth, U. (2006): *Innovationskultur in Deutschland - Rahmenbedingungen der Wissenschafts- und Technologiepolitik.* Discussion paper (as at July 2006).

Wieland, Th. (2006): *Innovationskultur: theoretische und empirische Annäherungen an einen Begriff.* In: Reinhold Reith et al. (Eds.) Innovationskultur: Modelle – Indikatoren – Transfer. Innsbruck: StudienVerlag, pp. 21-38.

Willis, D. (1997): *Der Delphin im Schiffsbug: Wie Natur Technik inspiriert.* Birkhäuser, Basel et al.

Winquist, F.; Krantz-Rülcker, C.; Lundström, I. (2004): *Electronic tongues.* In: MRS BULL, Vol. 29(10), pp. 726-731.

WWF [World Wide Fund for Nature] (Ed.) (1991): *Bionik: Patente der Natur.* [chief editor R. Witt; C.-P. Lieckfeld]. Pro-Futura-Verlag, München.

Zeng, F. (2004): *Trends in Chochlear Implants.* Trends in Amplification, Vol. 8(1).

Zerbst, E. W. (1987): *Bionik: Biologische Funktionsprinzipien und ihre technischen Anwendungen.* [Teubner Studienbücher der Biologie]. B. G. Teubner, Stuttgart.

Zeuch, M. (2006): *Bionik.* [Was ist was? Vol. 122]. Tessloff, Nürnberg.

8 ANNEX

8 ANNEX

8.1 List of illustrations

8.2 List of tables

MIX
Papier aus verantwortungsvollen Quellen
Paper from responsible sources
FSC® C105338

If you have any concerns about our products,
you can contact us on
ProductSafety@springernature.com

In case Publisher is established outside the EU,
the EU authorized representative is:
Springer Nature Customer Service Center GmbH
Europaplatz 3, 69115 Heidelberg, Germany

Printed by Libri Plureos GmbH
in Hamburg, Germany